Routledge Revivals

Ecological Modeling

This volume, originally published in 1975, grew out of Resources for the Future's involvement as a consultant to the Marine Ecosystem Analysis programme management within the National Oceanic and Atmospheric Agency. Here, researchers look at the state of the art in aquatic ecological modelling in a resource management context. Although the aim of the research in this volume is specific, the models used can be applied in broader contexts and provide conceptual frameworks for regional residuals-environmental quality management and other ecological modelling. This title is suitable for students interested in Environmental Studies.

Ecological Modeling

In a Resource Management Framework

Edited by
Clifford S. Russell

RFF PRESS
RESOURCES FOR THE FUTURE

First published in 1975
by Resources for the Future, Inc.

This edition first published in 2016 by Routledge
2 Park Square, Milton Park, Abingdon, Oxon, OX14 4RN
and by Routledge
711 Third Avenue, New York, NY 10017

Routledge is an imprint of the Taylor & Francis Group, an informa business

© 1975, Resources for the future, Inc.

Publisher's Note
The publisher has gone to great lengths to ensure the quality of this reprint but points out that some imperfections in the original copies may be apparent.

Disclaimer
The publisher has made every effort to trace copyright holders and welcomes correspondence from those they have been unable to contact.

A Library of Congress record exists under LC control number: 75015108

ISBN 13: 978-1-138-10092-3 (hbk)
ISBN 13: 978-1-315-65714-1 (ebk)
ISBN 13: 978-1-138-10118-0 (pbk)

ECOLOGICAL MODELING

In A Resource Management Framework

Clifford S. Russell, *Editor*

The Proceedings of a Symposium
Sponsored by National Oceanic
and Atmospheric Administration
and Resources for the Future

Resources for the Future, Inc.
Washington, D. C.

July 1975

This material has been published as received without the usual editing and typesetting in order to speed its distribution.

Library of Congress Catalog Card Number 75-15108
ISBN 0-8018-1773-0

Manufactured in the United States of America

Originally published, 1975
Second printing, 1976

RFF Working Paper QE-1 $6.00

LIST OF PARTICIPANTS

Dr. Carl Chen
Tetra Tech, Inc.
Lafayetteville, California

Dr. Nicholas L. Clesceri, Director
Rensselaer Fresh Water Institute at
   Lake George
Rensselaer Polytechnic Institute
Troy, New York

Mr. Leonard T. Crook
Executive Director
Great Lakes Basin Commission
Ann Arbor, Michigan

Professor Dominic M. DiToro
Civil Engineering Department
Manhattan College
Riverdale, New York

Dr. Howard Harris
National Marine Fisheries Service
Northwest Fisheries Center
Seattle, Washington

Dr. Robert Harriss
National Science Foundation
Washington, D.C.

Dr. Robert A. Kelly
Resources for the Future, Inc.
Washington, D.C.

Dr. Allen Kneese
Department of Economics
University of New Mexico
Albuquerque, New Mexico

Dr. Robert T. Lackey
Department of Fisheries and
   Wildlife Sciences
Virginia Polytechnic Institute
Blacksburg, Virginia

Dr. William Lavelle
Physical Oceanography Laboratory
Atlantic Oceanographic and
   Meteorological Laboratories
Miami, Florida

Dr. Thomas Murray
SEA GRANT Association
Washington, D.C.

Dr. Jacques C. L. Nihoul
Director of the Belgian National
   Environment Program - Sea Project
and Professor at the Universities
   of Liege and Louvain
Liege, Belgium

Professor Donald O'Connor
Civil Engineering Department
Manhattan College
Bronx, New York

Dr. Joel S. O'Connor
MESA New York Bight Project Office
Marine Science Research Center
Stony Brook, New York

Dr. Robert V. O'Neill
Ecological Sciences Division
Oak Ridge National Laboratories
Oak Ridge, Tennessee

Dr. Gerald T. Orlob, Principal
Resource Management Associates
Lafayette, California

Ms. Alician V. Quinlan
Ralph M. Parsons Laboratory
Department of Civil Engineering
Massachusetts Institute of
   Technology
Cambridge, Massachusetts

iii

Dr. Andrew Robertson
U.S. Department of Commerce
National Oceanic and Atmospheric
  Administration
Ann Arbor, Michigan

Dr. Clifford S. Russell, Director
Regional and Urban Studies
Resources for the Future, Inc.
Washington, D.C.

Mr. Donald Scavia
Rensselaer Fresh Water Institute at
  Lake George
Rensselaer Polytechnic Institute
Troy, New York

Dr. William E. Schaaf
Atlantic Estuarine Fisheries Center
Beaufort, North Carolinia

Dr. Walter O. Spofford, Jr., Director
Quality of the Environment Program
Resources for the Future, Inc.
Washington, D.C.

Professor Robert V. Thomann
Civil Engineering Department
Manhattan College
Bronx, New York

Dr. James P. Thomas
National Marine Fisheries Service
Atlantic Coastal Fisheries Center
Highlands, New Jersey

Dr. Douglas A. Wolfe
National Marine Fisheries Service
Atlantic Estuarine Fisheries Center
Beaufort, North Carolina

TABLE OF CONTENTS

Table of Contents (continued)

vi

LIST OF FIGURES

List of Figures (continued)

A Discussion of CLEAN, The Aquatic Model of the Eastern
Deciduous Forest Biome

The Delaware Estuary Model

List of Figures (continued)

List of Figures (continued)

Fish Population Models: Potential and Actual Links
to Ecological Models

Fisheries and Ecological Models in Fisheries Resource Management

List of Figures (continued)

Present Problems and Future Prospects of Ecologic Modeling

Introduction
by
C. S. Russell

This volume and the symposium on which it is based grew out of Resources for the Future's informal involvement as a consultant to the Marine Ecosystem Analysis (MESA) program management within the National Oceanic and Atmospheric Agency (NOAA). During the early part of the conceptual modeling phase for the New York Bight project, we proposed to NOAA, with the encouragement of Dr. Allan Hirsch who then headed the MESA program, that a look at the state of the art in aquatic ecological modeling in a resource management context would be useful for the MESA staff. The formal proposal for a cost-sharing contract to cover the symposium and publication of the papers and some discussion was submitted in the winter of 1974. The contract was approved and signed in the early summer and the symposium was held as scheduled on 10-12 September 1974, at the Brookings Institution in Washington.

In organizing the symposium we had very much in mind the under-lined words in the first paragraph: "in a resource management context." We tried hard to commission authors and discussants who were working in the context of resource management. In some cases this meant that their work was done explicitly for management agencies. In others, the work would be classified as research but was being carried on within a conceptual framework aimed at developing information and tools which could be useful to management at some time in the future.

We did not attempt to explore the frontiers of aquatic ecological modeling per se--and this has led to some misunderstanding. Our intention was to concentrate the limited time on contributions which already take some account of the problems and limitations implied by the larger resource management framework.

Notes on Content

Walter Spofford, in the symposium's first paper, sets out very briefly the reasons for the existence of a resource management problem at all--that is, a problem requiring some intervention by public decision-making bodies. He also sketches the history of the development of water quality and ecological system models, and he then describes in general terms the use of such models for resource management purposes. In doing so he draws a simple, yet frequently ignored or forgotten distinction between environmental models, "which describe the impact on the natural world of exogenous inputs, such as residuals discharges, but where an economic criterion and subsequent ranking of management alternatives [public policies] is not implied."; and management models "which involve the ranking of sets of management options according to a given economic criterion." Environmental models are generally one of the key components of resource management models.[1]

---

[1] This is not to say that all the outputs of an environmental model, for example DO concentrations, must or even can be valued in money terms. The most frequent approach is to explore the costs on meeting alternative sets of constraints on these outputs.

A substantial part of Spofford's paper is then devoted to describing and formalizing the management framework we have found useful in our work at RFF and to discussing the computational implications of various types of environmental models within that framework. Finally, he suggests an ambitious set of questions for the symposium participants to address. These center on an honest appraisal of the state of the art of aquatic ecological system modeling; on current gaps and deficiencies; and on future prospects.

Spofford's exposition of a framework for management-relevant modeling set the stage for the discussion and critique of the examples which make up the heart of the symposium. The first case study, presented by Park, Clesceri and Scavia, describes the Lake George Models (CLEAN and CLEANER) which have grown out of the International Biological Program, Eastern Deciduous Forest Biome activities funded by the National Science Foundation. It is the conception and product of a group of scientists doing research on the Lake, and in its current form it is perhaps best characterized as a research tool. The group is, however, interested in the application of their model to management of Lake George's quality, and to that end they are in the process of adding to it other modules describing relevant features of the system. The model is marked by fairly complex biological representation and the extreme simplicity of its assumptions about mixing and transport phenomena. Thus, the ecological model has 29 compartments within which are differentiated, for example, 3 types of zooplankton and at least two types of fish. On the other hand, the entire lake is represented by a single column of water.

The additional modules intended to create a management module involve the effects of tourist activity on nutrient loadings entering the Lake, an exogenous input to the ecological model. The group is apparently quite satisfied with the model's performance in verification runs and feels that model, output including, for example, predictions of lake trout catch rates, can be taken seriously by public policy makers and the public at large.

The second model described here is that of the Delaware Estuary produced by Kelly to be imbedded in the regional environmental quality management model developed at Resources for the Future. Kelly is a biologist, and his model reflects his concentration on biological processes with rather simple assumptions being made about the other facets of the problem. In contrast to the Lake George effort, however, Kelly was asked to develop a model reflecting existing techniques and data and was given no opportunity to play the model off against data gathering. In addition the characteristics of the exogenous inputs (residuals, or pollutants in the older parlance) were to some extent predefined by the data available on the dischargers along the estuary. Finally, Kelly had to work within a computational constraint imposed by the regional optimization framework. There had to be considerable attention given to the time required to find a steady state solution of the ecological model, and at the same time it had to be possible to calculate the matrix of first partial derivatives connecting endogenous and exogenous variables for every solution. The result is an

11-compartment ecological model for each of 22 reaches of the Delaware Estuary. The complex tidal hydrology is reduced to simple net downstream advection though Kelly does show results calculated with upstream dispersion taken into account. While the results are quite good when the model is run with low stream flow, the situation for which it was developed, they are very bad when a high stream flow is used, suggesting that one must treat the output of the model under all conditions with extreme caution.

The large international study being undertaken in the southern bight of the North Sea was discussed by Professor Nihoul of the University of Liége, and currently chairman of the Joint North Sea Modeling Group and Director of the Belgian National Environmental Program-Sea Project. One is particularly struck by two contrasts between this effort and the first two described. First, it is clear that hydrodynamicists and chemists have had a great deal to say about the project from the beginning, since, for example, the submodel of the currents in the bight is a complicated work in itself. Indeed, the biological modeling at this point seems to be the least developed part of the overall machinery. The second arresting fact is that this model is already being used in decision making -- in particular in the review of requests for permits to dump various wastes into the bight. Thus, even as data gathering, modeling, calibration and verification proceed, the semi-finished product is being consulted by government agencies. Nihoul's paper does not provide much detail, but he does make it clear that the focus of the ecological modeling at this point is the prediction of concentrations of heavy metals in fish, the benthos and other elements of the system resulting from particular rates of discharge from rivers and ocean outfalls and dumping from barges.

The final paper in the sequence describing specific models really deals with a number of individual models developed by Di Toro, O'Connor, and Thomann in the course of their consulting work. Thus, there are several levels of complexity described, on both the biologic and hydrologic sides. In general, however, the work of this group illustrates a route one might characterize as the complication of dissolved oxygen models, while the other three efforts, particularly the first two, represent simplifications of the enormously complex, but generally implicit, models biologists have of the aquatic systems with which they deal. It is also true that the models described in this paper have all been developed for management agencies and most of them have actually been used in decision making. In fact the authors are quite emphatic in telling us that they are not building research tools but instead are trying to use the existing fruits of research to put together working tools -- and the simplest possible tools for the given jobs.

Specifically, O'Connor, et al. describe applications of a general water quality model reflecting trophic-level interactions up through the zooplankton. The first application involves prediction of the effect of water diversions and increased nutrient discharges (due, in turn, to increasing population and economic activity) on phytoplankton populations in the lower American River. Estimates of the effect of alternative policies regarding both diversions and required sewage treament levels are included.

A second example of application, also involving a river estuary, is the case of the Potomac below Washington. Here the major management problem is the discharge of the nutrients from the Blue Plains Treatment plant and the resulting algal blooms which seriously depress the value of the estuary for recreational use. But the authors show that even very stiff requirements for nutrient removal at the plant may not clear up the algae, since very large amounts of nutrients enter the estuary in the river itself. These, being due largely to dispersed sources such as agricultural runoff, are not so straightforwardly controlled.

The third and fourth applications are to the problem of large lakes - specifically Erie and Ontario. These demonstrate the complex spatial structure necessary to capture the situation in water bodies with wide variations in depth, a lesson directly applicable to the N.Y. Bight study area. The Lake Ontario study further illustrates the difficulties for management policy making implied by long flushing times and stable nutrient exchange and storage systems. Altogether, the paper provides a strong demonstration of the utility of aquatic ecological models in the water quality management context.

Thus, the model applications represent wide variations along several dimensions, especially in terms of inception, intention and complexity. The reader, particularly the reader who is not already an expert will, however, be struck by the very great conceptual and structural similarities among the models. It is especially interesting to see how close to each other the biologists and engineers come out when the one group works to simplify its models enough to obtain practically computable structures, and the other departs from the classic, highly simplified and aggregated models of oxygen decay.

We commissioned two other papers to deal with specific topics which we anticipated would not come up in the case studies but which would be of interest to the MESA program staff -- and to others as well, we hope. First, we asked William Schaaf to discuss the prospects for linking traditional fisheries models with aquatic ecosystem models. Our guess here is that such a combination of models may be the most efficient way of addressing questions about the impact of residuals discharge and other acts of man, on fisheries. Although nothing definitive falls out of the symposium on this subject, there was general agreement that existing ecosystem models are not reliable as predictors of fish populations and that existing fisheries models are not tailored to handle pollution problems -- so the need exists for new and better linkage.

We also asked Robert O'Neill to give us the benefit of his experience as modeling coordinator for the Eastern Deciduous Forest Biome part of the IBP by discussing the management of modelers and large multi-sited modeling enterprises. This topic is likely to become more and more important as large scale, truly interdisciplinary modeling efforts are mounted in areas of national concern, such as the effects of energy resource development on terrestial and aquatic ecological systems and hence on the quality of human life. It is already relevant to MESA, which involves a number of physically separate facilities of NOAA, and within those facilities, scientists from several disciplines. O'Neill's remarks deal specifically with the difficulties raised by mathematical modeling, where that activity is undertaken by individuals and groups distinct from the data gathering and analyzing participants. This again is likely to be a particular problem for the MESA effort.

In the final paper, Orlob provides the symposium with its cap-stone in the form of a summary and analysis of "Present Problems and Future Prospects for Ecologic Modeling." This paper, based both on the literature generally and on Orlob's reading of the conclusions of the symposium, deals with a very broad spectrum of topics. Orlob begins with a reprise of Spofford's introduction, discussing the management context in general terms. He goes on to deal with two questions which seemed particularly to trouble the participants: who are the modelers and who are the decision makers? Are they in fact usefully seen as distinct groups? (Orlob thinks they are.) He further discusses the process of model development, identifying 6 stages, beginning with conceptualization, running through functional representation, computational representation, calibration, verifica-tion and ending with documentation. All these steps, of course, he sees as prior to application, the presumed goal.

A key section of Orlob's paper is an assessment of the present status of ecological modeling. Here he discusses the situation with respect to: data; hydrodynamics, hydrology and the ecosystem; ecological concepts; aquaculture: acute vs. chronic effects; prediction, evaluation and reliability; adaptability and transferability: and bacteria. It would hardly be appropriate for me in this introduction to anticipate Orlob's observations and conclusions in these areas. Rather, I shall confine myself to four general remarks which I think describe the general tenor of Orlob's paper and of the symposium itself.

1. For reasons of data, conceptual, and computational limitations, aquatic ecological models must currently be viewed as reliable predictors of system behavior only up to the level of phytoplankton biomass. In particular, current models do not seem to be able to deal adequately with fish stocks.

2. There is a need for generally accepted verification procedures.

3. Certain specific system elements, now often ignored, should be included in existing models: for example, bacteria and the anaerobic cycle.

4. More attention should be paid to dynamic behavior of aquatic systems under stresses. In particular, models should be developed which allow investigation of collapse probabilities. As a conceptually related matter, methods should be developed for predicting effects of chronic, sub-lethal stresses.

The reader is not forced to take my word that these remarks capture anything like the sense of the meeting. We have included, as an appendix, a transcript of the discussion surrounding the papers. This has been edited to eliminate repetition and to improve on the mode of expression. But no substantial topic or point of view has been stricken.

I speak for my colleagues at RFF in expressing the hope that this volume will be of some value to NOAA/MESA and to the larger modeling community as well.

* * *

This volume reflects the efforts of several of my colleagues at RFF. Allen Kneese, now at the University of New Mexico, made the initial contacts and helped us to get the idea off the ground. Walter Spofford and Robert Kelly assisted at every stage of the venture, suggesting contacts, reading papers, and generally making up for my lack of expertise in this area. Dee Stell took care of many of the details necessary to a smoothly running conference, and Brenda Lonning made a photo-ready manuscript out of the hodgepodge of individual papers initially received.

ECOLOGICAL MODELING IN A RESOURCE
MANAGEMENT FRAMEWORK: AN INTRODUCTION

Walter O. Spofford, Jr.*

I. Introduction

Virtually all of man's production and consumption activities have
an impact on the natural systems of the world. Some of these activities
have resulted in substantial changes to the aquatic and terrestrial environments, while others have resulted in changes that are hardly per-
ceivable. At one extreme, these changes have been temporary and confined to relatively localized areas as in the case of discharges of organic materials to rivers and streams, and where these natural systems
have recovered fairly rapidly to their original states after the residuals (waste) discharges were stopped. At the other extreme, the environmental changes have been widespread and long-term, and even more permanent, as in the case of discharges of mercury and DDT to the environment which already may have resulted in irreversible changes (however
small) and which may cause changes in the genetic structure, or even
result in the extinction, of certain species.

Man simply cannot exist at current population levels

* The time involved in the preparation of this paper while a member
of the research staff at the International Institute for Applied Systems Analysis in Vienna, Austria, is gratefully acknowledged.

and densities without causing some perceptible changes to natural
systems, both locally and globally. And as the living standards increase throughout the world, and more and more countries convert from
agricultural to industrial economies, the more likely are these environmental changes to be substantial. (There are, of course, many counter-
examples where some of the less developed nations have caused disastrous
changes to their environments through poor management practices. Over-
grazing, over-cutting of timber resources, and the use of poor agricultural practices that have resulted in excessive erosion are some examples).
Choices ultimately must be made between economic, or material prosperity,
levels on the one hand and different types of natural environments, sometimes referred to broadly as levels of "environmental quality", on the
other. This is what the management of environmental resources is all
about. One of the challenges we face as a society is how to achieve a
higher standard of living (however defined) and at the same time maintain
a decent quality environment, both now and in the future.

Public management of environmental resources

Two of the more important issues currently facing society regarding the uses of environmental resources are the selection of an appropriate organizational structure for managing these resources and the
selection of a process for allocating them among competing uses. There
are at least two justifications for governmental, or public, management
of most environmental resources, especially of the ecosystems: (1) they
exhibit characteristics of public goods; and/or (2) they are common
property resources.

The essential characteristics of a public good (or "bad") are the following: the consumption of it is the same for all members of society; it can not be provided without everyone receiving the same level; and the consumption by one member of society does not affect the consumption by others. Consequently, where public goods are involved, there is a need for a collective choice mechanism, or process, which the government can provide, to select socially desirable levels. Examples of public goods are national security, and air and water quality.

The main feature of a common property resource is that it cannot, for physical or institutional reasons, be owned privately. Because of this, the use of the resource is not allocated by the private market system, and hence there is an incentive for individuals to act as nonowners and overuse the resource and to impose costs and damages on others. In this case, government, or public, intervention is required to prevent overuse and to prevent undesirable distributions of costs and damages from being imposed on members of society. Degradation of the environment is an excellent example of the failure of the private market to allocate efficiently a common property resource. Examples of common property resources are the air mantle, waterways, aquatic and terrestrial ecosystems, and fresh water fisheries in this country.

Not all public goods are common property resources, and vice versa. For example, fresh water fisheries in this country are not public goods, but they are common property resources. The public good characteristics, and the common property nature, of environmental resources represent two, quite independent, justifications for public management of these resources.

Given that public management of most of our environmental resources is necessary, we are now faced with the following questions: What is an appropriate social choice mechanism for making collective choices on the uses of our environmental resources and what institutional arrangements --existing or otherwise--are required for this purpose? Perhaps the most difficult problem and one furthest from solution in the management of common property resources is how to make value choices involving their use. This will remain a challenge for social scientists, and society generally (Russell and Kneese, 1973). It is not the purpose of this _Introduction_, or of the Symposium, however, to explore the social choice matter in detail; only to mention it in passing. This aspect of the management of our environmental resources has been, and continues to be, given extensive attention by others.[1] Rather the purpose of this _Introduction_ is to provide a framework for the analysis of environmental resource management problems that incorporate water quality models, and aquatic ecosystem models, as essential components.

The development of water quality and ecosystem models

The employment of water quality models for management purposes is not new. Ever since Streeter and Phelps first presented, in 1925, their well-known formulation for predicting the dissolved oxygen levels of rivers, water quality models, in one form or another, have found much use in the management of water resources (Streeter and Phelps, 1925). Over the years, various attempts have been made to improve the predictive capability of these models. Streeter and Phelps' original formulation included a description of the carbonaceous oxidation process (first-

---

[1] See, in particular, Haefele (1973, 1975); Russell, Spofford, and Haefele (1974).

stage BOD) only. They did not include the effects on dissolved oxygen of nitrogenous oxidation (second-stage BOD); production of oxygen through algal photosynthetic activity; reduction of oxygen through biological and chemical oxidation (for example, plant and animal respiration causing benthal demands); or the effects of runoff, scour, sedimentation, and so on. In addition, their formulation was deterministic and steady state (with respect to the concentration at a particular point in space and time), and first-order kinetics for the biochemical reactions was assumed.

Other water quality modelers[1] have subsequently added some refinements in their treatment of the dissolved oxygen relationship in streams. But the approach has basically remained an empirical one extending the original formulation of Streeter and Phelps through the addition of new terms to the basic equations. Only two differential equations (three when nitrogenous oxidation is treated endogenously) are employed for this approach; the first one for describing the spatial and temporal distribution of organic material, the second one for dissolved oxygen levels. All factors other than the oxidation (decay) of organic material that affect the oxygen balance, and which are included explicitly, such as photosynthetic activity, are treated as exogenous parameters rather than as endogenous variables. And, as before, first order kinectics for all reactions is assumed.

More recently, however, because of the increasing relative impact of nitrogenous oxidation (nitrification) on the dissolved oxygen balance

---

[1] See, for example, Camp (1963) and Dobbins (1964). For an excellent summary of the DO relationship in streams, see O'Connor (1967). For stochastic formulations, see Custer and Krutchkoff (1969) and EPA (1971).

of water bodies, attempts have been made to model these reactions explicitly. Although the desired output remained the dissolved oxygen
levels, this represents a significant departure from the original
Streeter-Phelps model structure as no longer are organic material and
dissolved oxygen the only materials explicitly accounted for. The various nitrogenous forms--organic nitrogen, ammonia, and nitrite and nitrate ions--are considered as well.[1] With this approach, one equation
for each endogenous variable is required for simultaneous solution.

Even more recently, because of the public interest in eliminating
massive algal blooms and eutrophication problems from our waterways,
attention has shifted, slightly, away from the prediction of DO levels
toward the prediction of algal densities. Consequently, in the past
four or five years, considerable effort has gone into the development
of phytoplankton models (see, for example, DiToro et al., 1970; and
Chen, 1970). Some of these phytoplankton models are intended to be used
for predicting DO levels as well as algal densities, whereas others are
not. And the treatment of the relationship between algae and dissolved
oxygen, if the latter is included at all, varies among modelers.

Up to this point in time, the motivation for the majority of water
quality modeling efforts has been to improve the capability for predicting, in both space and time, the distribution of dissolved oxygen in
water bodies.[2] As pointed out above, some effort has also gone into

---

[1] For an example of a multi-stage system for describing the nitrification phenomenon, see Thomann et al (1970) and Manhattan College (1970).

[2] It should be pointed out, however, that much effort has also gone
into the development of water quality models for predicting the concentration distribution of conservative (for example, salinity) as well as
other nonconservative (e.g., radioactive materials and heat) residuals.

the development of models for predicting algal densities. The major effort in the development of these models, especially of the oxygen prediction type, has been by engineers. The use of these models in a management context has been the motivating force behind their development.

While engineers were developing their rather pragmatic water quality models for management purposes, ecologists were developing their own set of models for a variety of purposes, including research as well as management.[1] Interests in, and approaches to, the development of the latter set of models differed substantially from those of the water quality modelers. Ecologists working in the field of aquatic ecosystem modeling traditionally have been interested in species populations at various trophic levels, and to them dissolved oxygen and even nutrients, while perhaps important components of the model (and system), are often considered secondary outputs. The biological aspects of their models are generally more complex than the ones discussed above, and their models are based more on biological mechanisms and less on empiricism. Feedforward and feedback relationships among biological components are included explicitly, but the forms of these relationships vary considerable among ecosystem modelers. The tradeoff among what and how many species and materials to account for explicitly, and available computer size and budget, remains a problem with them. This tradeoff is related directly to the ultimate use of the model--research, management, or other. This, in itself, requires a great deal of experience as in most cases this choice is as much an art as a science. The development of

----

[1] For a review of early plankton production models, see Patten (1968). For a discussion of a typical simulation model of an aquatic environment, see Parker (1968).

the majority of these models to date has been motivated primarily by research interests. Fisheries models are an exception to this, as are those developed for examining the impact on the aquatic environment of residuals (waste) discharges.[1]

The mathematical forms of all the models discussed above are similar. They are all made up of sets of differential equations with one equation for each material or species accounted for explicitly within the model. And in general, these equations must be solved simultaneously, although not always.[2] These models are generally referred to as "compartmental" models by systems ecologists. The Streeter-Phelps model would be considered a two-compartmental model--one compartment for organic material, the other for dissolved oxygen. A typical eleven-compartmental model presented later on in this volume by Robert Kelly considers the following components: algae, zooplankton, bacteria, fish, dissolved oxygen, organic matter (as BOD), nitrogen, phosphorus, toxics, suspended solids, and temperature.

It is interesting, and certainly encouraging, to note that even though some multi-compartmental models are being developed by one group of modelers primarily for the purpose of predicting dissolved oxygen

---

[1] Examples of ecosystem models developed for residuals management purposed include: The Delaware Estuary (Kelly, forthcoming), Narragansett Bay (Nixon and Kremer, forthcoming), and the Baltic Sea (Sjöberg et al, 1972, and Jansson, 1972).

[2] For example, if biological and/or chemical reactions do not contribute substantial quantities of heat energy to the body of water (which they generally do not), the temperature distribution can be evaluated separately and treated as an exogenous input to the remaining portion of the model (see, for example, Bloom et al, 1969).

levels and algal densities, while others are being developed by another group of modelers primarily for predicting species population levels, the models developed specifically for management purposes are becoming more and more similar. The major differences between the two modeling approaches now appear to be in the way the physical transport and mixing phenomena (hydrodynamics) are incorporated, and in the number, and designation, of the compartments in the model and the forms of the feedforward and feedback relationships among them. But some of these differences would (and do) occur even among ecologists, as at the present time there is no universal agreement on how all the different biological mechanisms should be represented mathematically.

The use of water quality and ecosystem models for management purposes

For aquatic ecosystem models to be useful in a management context, they must be able to accept as inputs man-induced changes and disturbances, such as modified hydrological regimes or the introduction of residuals (waste) discharges, and they must provide, as output, information that is relevant and meaningful for making policy decisions on the level of use of a given water resource. The input and output information will not be dictated by the ecosystem modeler alone, but by the more encompassing resources management system within which the ecosystem model forms an important part. For residuals management and environmental quality analyses, the most significant inputs to the aquatic ecosystem are residuals discharges from industrial activities and municipal sewage treatment plants, and runoff from urban, agricultural, and forest management activities. The ecosystem model should be structured to accept these residuals as inputs. Examples of ecosystem model outputs of interest to people and policy makers are algae concentrations (aesthetic considerations), fish populations (recreation and commercial considerations), pathogenic bacterial counts (recreational and health considerations), and even dissolved oxygen levels because of their importance as an indicator of the general health of the system. The policy makers most likely will not be interested in the ambient concentrations of inorganic phosphorus or nitrogen, although this may be extremely important to the ecologist and to the subsequent behavior of the ecological system.

The variables that describe the behavior of the ecological system are referred to as state variables. Their levels at any point in time and space are a function both of the other state variables and of exogenous influences. Some of the ecosystem state variables are of interest to policy makers, especially those that best convey to the layman the state of the natural world. When socially desirable upper, or lower, limits for these variables are set in the political process, they are known as policy targets and commonly referred to as ambient standards.

The management, or policy, variables form an essential part of the management model. The levels of these variables can be controlled by society, though generally at a cost. These variables, in turn, influence the behavior of the natural systems through changes in inputs. Policy instruments are used to affect the levels of these inputs. Policy instruments are the economic, legislative, executive, and legal means we have for influencing the levels of the policy variables. Effluent standards, effluent charges, taxes, and restrictions on the use of materials or on the production of final products are examples of policy instruments.

For management purposes, ecosystem models can be employed in two basic ways. First, they can be used, by themselves, as tools for predicting alternative states of the natural world, <u>given</u> various sets of residuals input levels and distributions. This mode of operation is commonly referred to as the scenario, or simulation, mode. It represents a purely descriptive use of the model. Second, ecosystem models can be used in conjunction with optimization, or management, models with "optimal" levels and distributions of residuals discharges (inputs to the ecosystem) selected by the management model, <u>given</u> ambient standards. This use of the ecosystem model, in conjunction with economic cost information and ambient standards or environmental damage functions, represents a normative, or prescriptive, use of the model. Unless the aquatic ecosystem model is linear (in responses and discharges), this use of the model is not as mathematically tractable as the former. The mathematical aspects of the latter use of aquatic ecosystem models will be presented in considerably more detail in section III of this paper.

<u>A preview of the following sections</u>

As we stated above, the purpose of this paper is to provide a framework for the analysis of environmental resource management problems that incorporate ecosystem models as an essential component. To this end, a regional environmental quality management system, including the range of management options available to society for maintaining, or improving, environmental quality, is described qualitatively and quantitatively in sections II and III. In addition, we indicate how all the components of the system, especially the ecosystem models, may be linked together within the same conceptual framework. In the last section, we present a summary, suggest alternative ways of, and considerations for, managing aquatic ecosystems, and finally end up with a set of questions for the symposium participants relating to the current state-of-the-art of ecosystem models developed for management purposes.

Before proceeding, however, we must define some terms. The use of the term _management model_ is restricted in this paper to those situations which involve the ranking of sets of management options according to a given economic criterion. Optimization (or programming) models come under this category, as do simulation models when costs and benefits associated with the various exogenously selected management alternatives are delineated, compared, and ranked. The term _environmental model_ is reserved for those models which describe the impact on the natural world of exogenous inputs, such as residuals discharges, but where an economic criterion and subsequent ranking of management alternatives is not implied. Environmental quality models, and in particular ecosystem models, are a necessary part of the more encompassing environmental resource, or environmental quality, management models.

II. The Regional Environmental Quality Management System

A regional environmental quality management model that includes all the relevant management options available to society for improving environmental quality, or for reducing the impact on the natural systems of the undesirable side effects of production and consumption activities, consists of the following components:

1. Final demand functions: provide a basis for estimating how the demand for final products will change under the influence of different pricing policies and restrictions on sale.

2. Residuals generation and discharge models: describe the factors influencing the generation, modification (e.g., treatment), and final discharge to the environment of residuals from production and consumption activities, including relevant costs. These models encompass all types of economic activities in the region within which residuals are generated and subsequently discharged to the environment. Examples of activities within this category are (1) industrial production, both of intermediate and final goods; (2) mining activities; (3) agricultural activities; (4) forest (management) activities; (5) urban and suburban runoff; and (6) consumption activities (both commercial and residential residuals generators) including municipal wastewater treatment facilities. Models of these activities range from comparatively simple statistical (regression) functions relating outputs and inputs to relatively sophisticated linear programming models (see Russell, 1971).

3. <u>Environmental modification models</u>: describe the options available for improving the assimilative capacity of the environment,
   including their costs. Provision of additional river flow
   through low-flow augmentation and the addition of dissolved
   oxygen to water bodies through in-stream aeration are examples
   in this category.

4. <u>Environmental quality models</u>: translate the time and spatial
   patterns of residuals discharges and the time and spatial patterns of natural influences such as sunlight and river flow
   into time and spatial patterns of the resulting states of the
   natural environment (described by ambient residuals concentrations and population sizes of biological species and/or functional groups of interest).

5. <u>Damage functions</u>: relate time and spatial patterns of ambient
   residuals concentrations to the resulting impacts on receptors--
   man, animals, plants, and structures--in physical, biological,
   and economic terms. Environmental standards are generally set
   with respect to the impact on receptors in physical and biological terms. Economic damage functions are seldom available.

6. <u>Management strategies</u>: consists of alternative sets of measures that affect one or more points in the management system,
   together with the benefits, costs, and damages associated with
   each strategy.

The conceptual framework

A conceptual framework for regional residuals-environmental quality management is depicted in Figure 1. Starting at the upper left-hand corner of this figure, a spatial distribution of economic activities is selected (or assumed). A specified vector of final demand (which can be changed) and a set of alternative production processes and treatment activities result in the discharges of residuals to the environment. These discharges are treated as exogenous inputs to the environmental quality (or ecosystem) models. The ecosystem may also be affected by certain modifications such as increased flow through low-flow augmentation and additional oxygen through in-stream aeration. The combination of residuals inputs, man-induced environmental modifications, and natural conditions (storms, sunlight, etc.) results in a modified natural environment. These changes to the environment are generally measurable in physical, chemical, and/or biological terms; and sometimes, in economic terms. For management purposes, these environmental changes are often compared with ambient standards that are set in the political process. If the environmental damages are excessive (in the sense of being socially undesirable), or if the ambient standards are exceeded, a new management strategy can be selected to change the mix of final demand; change the amounts, and time and spatial patterns, of residuals discharges; increase the levels of environmental modification activities; or if all of the above are simply too costly from society's point of view, the ambient standards can be raised or lowered as desired.

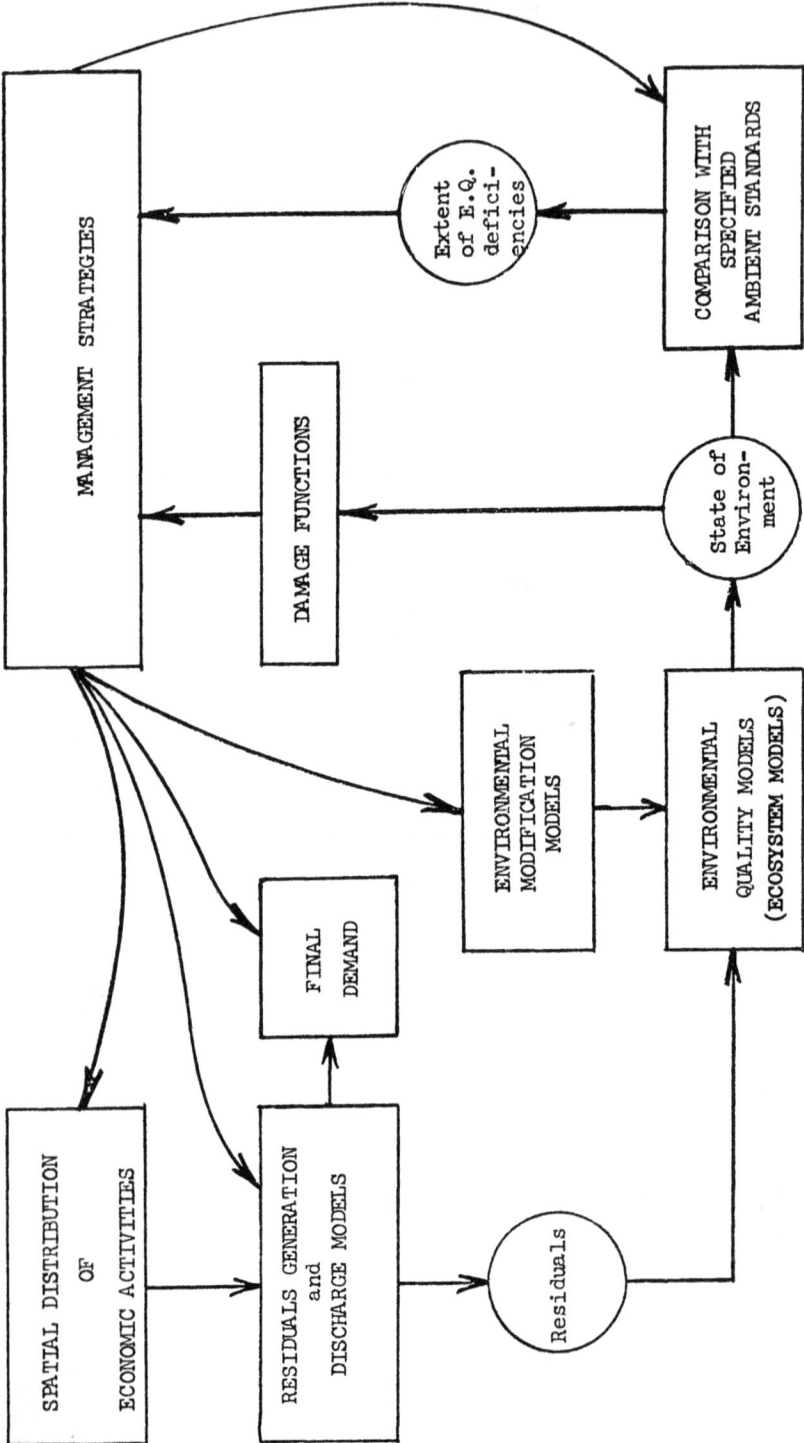

Figure 1. SIMPLIFIED SCHEMA OF CONCEPTUAL FRAMEWORK FOR RESIDUALS-ENVIRONMENTAL QUALITY MANAGEMENT

Management strategies

A management strategy consists of a set of measures or activities affecting the vector of final demand, one or more of the residuals generators, the natural environment, or the receptors. These strategies can be evaluated in the context of different objective functions and different sets of constraints including both effluent and ambient environmental quality standards. The mere act of ranking strategies implies a prescriptive approach to the management of environmental resources. Basically, there are three approaches to seeking an optimal strategy for any given objective and set of constraints: (1) response surface sampling using simulation, (2) optimization (mathematical programming), and (3) a combination of (1) and (2). An example of the latter is exogenous treatment of various levels of low-flow augmentation in a water quality optimization model. However, the costs (and damages if they occur) of providing the various augmented flows are included in the overall ranking of the various management alternatives.

Before proceeding with a discussion of the advantages and disadvantages of each approach, it might be helpful to be a little more explicit in what we mean by the term simulation model. It is used in two different ways in this paper. One use is associated with management models and the ranking of management alternatives (prescriptive use); the other, with environmental, or ecosystem, models (descriptive use). In the first case, the simulation output we are interested in is the ranking criterion (costs, benefits, or net benefits). This use may be characterized by the (objective) function

$$C = f(X)$$

where

> C = total discounted costs, benefits, or net benefits
>
> X = vector of management alternatives and system state variables.

In the second case, the simulation output we are interested in is the state of the natural system in physical, chemical, and/or biological terms. This simulation may be expressed as,

$$S(x,t) = f[B,Z(x,t),J(x,t)]$$

where

> $S(x,t)$ = the state variables of the system specified in terms of space and time coordinates
>
> B = vector of system parameters
>
> $Z(x,t)$ = man-induced forcing functions (such as residuals discharges)
>
> $J(x,t)$ = natural forcing functions (such as storms, sunlight, etc.).

Simulation models, in general, are able to provide a more realistic representation of real world conditions, and their outputs are generally easier to obtain than optimization models are to solve. They are conceptually straightforward, and nonlinearities, discontinuous functions, nonsteady-state (transient) behavior, and stochastic aspects are much easier to include than with optimization models. Their major disadvantage is the general difficulty of selecting a priori that combination of economic activity levels that optimizes a given objective function. Exhaustive sampling of a finite number of combinations can be used. But because the total number of combinations is usually extremely large, random sampling techniques appear to be a more reasonable approach.

The major advantage of optimization models is the direct determination of the activity levels that optimize a given objective function. Their major disadvantage, given the magnitude of the regional resource management problem, is that they are generally difficult to construct and then to solve, even when formulated as linear programming problems. Furthermore, they may not be good representations of the actual (real world) situation. For some cases, a combination of simulation and optimization techniques provide the logical approach to environmental management problems. The use of one technique or the other, or a combination, would depend upon each individual situation.

III. Management Model Formulation

In this section, a formal mathematical description of a regional residuals-environmental quality management model is presented. This description clearly indicates where the ecosystem models fit within the more encompassing management model and also depicts the essential links between the ecosystem model and the other parts of the management system. The environmental management problem in this section is cast in terms of an optimization model. And for purposes of exposition, some assumptions are made.

The management model presented here is deterministic and steady state. Only one season, which could represent either the low-flow season or an entire year, is considered at a time. Steady-state values for the ecosystem state variables are used for comparison with ambient standards. From an economic point of view, the model is static; economic and population growth, and capacity expansion, are not considered explicitly. It is possible to extend this model to incorporate dynamic aspects. But this would unnecessarily complicate the exposition.

The type of objective function that is appropriate for analyzing regional environmental quality management problems depends upon the particular situation, data availability, and the way social choices are made on levels of environmental quality and costs of achieving them. This problem, however, will not be examined here. For purposes of exposition, a net benefit function to be maximized is assumed to exist (although a minimum cost objective function could have been dealt with just as easily).

In this discussion, the concentration will be on overall model formulation. The component parts will not be developed in detail. We will carefully examine the mathematical forms of the various environmental models, concentrating in particular on linear and nonlinear ecosystem models, and indicate how they might be handled within this framework. For this development, we assume that residuals generation and discharge models exist in standard linear programming form. (This assumption is not essential, however, but it does allow us to make a rather straight-forward presentation.)

Mathematical statement of management model

Let us now state the environmental quality management problem formally.

$$\max \left\{ F = f(X,R) \right\} \tag{1}$$

$$\text{s.t.} \quad AX \leq B \tag{2}$$

$$h_i(X) = R_i \qquad i = 1,\ldots,q \tag{3}$$

$$R_i \leq S_i \qquad i = 1,\ldots,q \tag{4}$$

$$R_i \geq 0 \qquad i = 1,\ldots,q \tag{5}$$

$$X \geq 0 \tag{6}$$

where $f(X,R)$ is the objective function; $AX \leq B$ is a set of linear constraints comprised of both the equality and inequality types; $h_i(X) = R_i$, $i = 1,\ldots,q$, represents a set of environmental functions (for example, an ecosystem model) which relate ambient concentrations of residuals (and/or population sizes of species), R, and residuals discharges, X; X is a vector of decision, or management, variables, including residuals discharges; R_i, $i = 1,\ldots,q$, is a vector of ambient concentration levels

34

of residuals and population sizes of species; and S_i, $i = 1,...,q$, is a
vector of ambient environmental quality standards (e.g., algal densities,
fish biomass, and dissolved oxygen). Note that the environmental relationships could have been written directly as,

$$h_i(X) \leq S_i \qquad i = 1,...,q.$$

However, we choose to deal explicitly with the variables R_i, $i = 1,...,q$,
here as they will be useful to us in a later development.

If the objective function, eq. (1), and the constraints, including
the environmental models, eqs. (2) and (3), are all linear, this environmental management problem reduces to a standard linear programming problem of the form[1]

$$\max \left\{ cX \right\} \tag{7}$$
$$\text{s.t.} \quad AX \leq B \tag{8}$$
$$X \geq 0 \tag{9}$$

where now the vector R is included in the vector X. The classical water
quality management problems are of this form (see Sobel, 1965).

On the other hand, if either the objective function, eq. (1), or
the constraint set, eqs. (2) and (3), are nonlinear, the problem is a
nonlinear programming problem and it is more difficult to solve.

Sometimes the environmental models are available in linear form.
But frequently they are only available as nonlinear simulation models
and are virtually impossible to deal with using traditional mathematical
programming techniques (except of course where they can be linearized

---

[1] Note that the constraint sets, eqs. (2) and (8), are not the same.
Eq. (8) includes both eqs. (2) and (3).

directly).[1]

Because it is generally easier to deal with a linear constraint
set than with a nonlinear one, we remove the environmental relationships
from the constraint set and deal with them in the objective function.
This modification of the problem requires the use of penalty functions
for exceeding the environmental quality standards, eq. (4).[2]

The new optimization problem may be stated formally as

$$\max \left\{ F = f(X) - P(X,S) \right\} \tag{10}$$

$$\text{s.t.} \quad AX \leq B \tag{2}$$

$$X \geq 0 \tag{6}$$

where

$$P(X,S) = \sum_{i=1}^{q} p_i \, [S_i, R_i = h_i(X)] \tag{11}$$

and where $p_i(S_i, R_i)$, $i = 1,\ldots,q$, are the penalty functions associated
with exceeding the environmental standards, S_i, $i = 1,\ldots,q$. For reasons
that we will see below, we require that the function $P(X,S)$ be continuous
and have continuous first derivatives. Note that the q environmental
model relationships, $h_i(X) = R_i$, $i = 1,\ldots,q$, have been used to eliminate

---

[1] In terms of the complexity involved in incorporating steady-state
environmental models within an optimization framework, we find it useful
to distinguish among four broad categories: (1) linear relationships
where ambient concentrations are expressed as explicit functions of residuals discharges, i.e., $R = AX$; (2) linear, implicit functions, i.e.,
$X = AR$ (note that this equation set can be rearranged by inverting the
matrix of coefficients, i.e., $R = A^{-1}X$); (3) nonlinear, explicit functions, $R = f(X)$; and (4) nonlinear, implicit functions, i.e., $X = f(R)$.

[2] The penalty function approach for eliminating constraints is not
new. It is a well-known technique. See, for example, Fiacco and McCormick (1968). For an evaluation of penalty function methods versus other
methods for dealing with nonlinear constraint sets in mathematical programming problems, see McCormick (1971).

the q element vector of ambient concentrations, R_i, $i = 1,...,q$, from the original objective function, eq. (1).

The new optimization problem, eqs. (10), (2), (6), and (11), may be solved using one of the available nonlinear programming algorithms. Whatever optimization technique is employed, we must be able to evaluate the total penalties, $P(X,S)$, for various sets of residuals discharges. This requires solving the relevant environmental models for a given discharge vector, X, to determine the resulting state of the natural world, and then comparing this state with the ambient environmental quality standards. In addition, gradient methods of nonlinear programming require that the vector of marginal penalties, $\dfrac{\partial P(X,S)}{\partial X}$, be evaluated for each state of the natural world which has been computed. Regardless of the optimization scheme employed, this analysis requires a side computation involving the environmental models, and this computation must be made at each step in the ascent procedure. An algorithm that we have found useful for our research on residuals management models, and which utilizes a standard linear program for climbing along the nonlinear response surface, is shown schematically in Figure 2.

Environmental response matrix

From eq. (11), the vector of marginal penalties may be expressed as

$$\frac{\partial P(X,S)}{\partial X_j} = \sum_{i=1}^{q} \frac{dp_i}{dR_i} \cdot \frac{\partial R_i}{\partial X_j} \qquad j = 1,...,n \qquad (12)$$

or in matrix notation as

$$\frac{\partial P(X,S)}{\partial X} = \left[\frac{\partial R}{\partial X}\right]^T \cdot \frac{dp}{dR} \qquad (13)$$

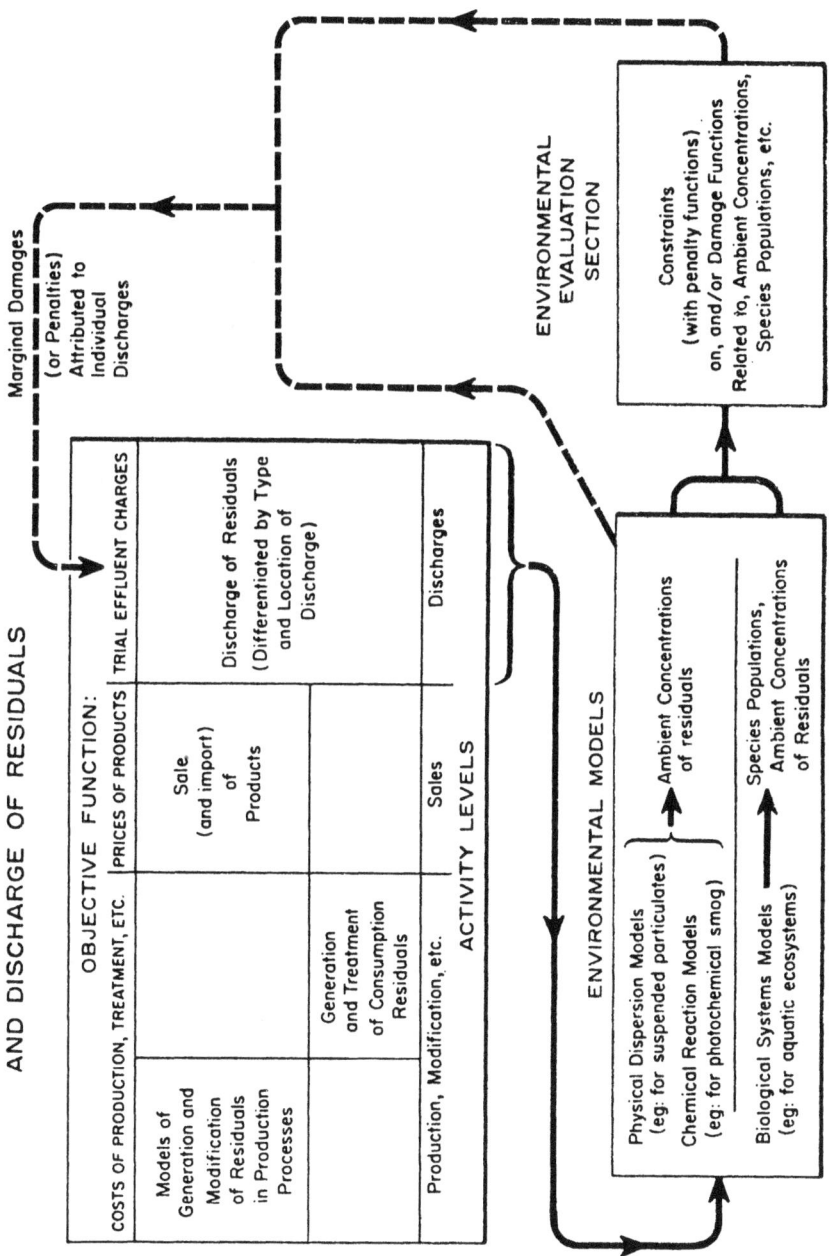

LINEAR PROGRAMMING MODEL OF REGIONAL GENERATION AND DISCHARGE OF RESIDUALS

FIGURE 2. SCHEMATIC DIAGRAM OF THE REGIONAL RESIDUALS MANAGEMENT MODEL

where $\left[\dfrac{\delta R}{\delta X}\right]$ is referred to as the "environmental response matrix", and $\dfrac{dp}{dR}$ is a vector of first derivatives, or slopes, of the individual penalty functions. The environmental response matrix provides us with a measure of the impact on the natural system of small (unit) changes in the inputs of residuals. For example, it yields the impact on the dissolved oxygen concentration in reach 20 of a unit change in the organic residual loading in reach 5. The elements of this matrix may be positive or negative.[1]

For linear environmental systems, $\left(\dfrac{\delta R}{\delta X}\right)$ is the matrix of transfer coefficients, A, when the environmental functions are expressed, linearly, as

$$R = h(X) = A \cdot X \qquad\qquad (3)$$

For nonlinear environmental models, the situation is similar except that the evaluation of the environmental response matrix $\left(\dfrac{\delta R}{\delta X}\right)$ is substantially more involved, and because the response surface is nonlinear, it must be recomputed for each state of the natural world.[2]

The management model development in this section was based on a steady-state ecosystem analysis. It was not intended to deal with the problem of peak concentrations associated with the nonsteady and stochastic aspects of water resource systems. This would have required a more

---

[1] For more information on the evaluation of the environmental response matrix for a nonlinear ecosystem model, and for ways that it might be used in water quality management, see Kelly and Spofford (forthcoming).

[2] For more detail on ways of incorporating nonlinear ecosystem models within an optimization framework, see Spofford, Russell, and Kelly (1975), and Spofford (1973).

sophisticated analysis and would be substantially more difficult to incorporate within an optimization framework. For example, the response matrix which currently involves only the space dimension, $\left[\frac{\partial R}{\partial X}\right]_x$, would have to be expanded to include the time dimension as well, $\left[\frac{\partial R}{\partial X}\right]_{x,t}$. However, as important as this consideration is for us, it is beyond the scope of the management model developed here. Ways of dealing with this problem are currently being examined by Professor François-Xavier de Donnea and his colleagues at the University of Louvain in Belgium.[1]

---

[1] This effort is being supported by a grant from Resources for the Future.

IV. Summary and Questions for Symposium Participants

The wise management of the natural systems of the world is one of the more imposing tasks facing society today. We have much to learn about the properties and behavior of natural systems, especially the properties of stability and resilience, and the domains of attraction of natural systems.[1/] Society's actions today can have a profound impact on the environment for many generations to come. And environmental insult in one area can have a far-reaching impact on natural systems great distances away. Indeed, if we are not extremely judicious in our actions, we could trigger a chain of events that could result in major, and irreversible, changes in the environment.

Most modeling efforts in ecology to date have concentrated on equilibrium conditions and states, and on small changes around these equilibrium states, using deterministic, nonsteady-state simulation models. Recently, because of the concern for environmental degradation and endangered species, attention has shifted away from equilibrium states to conditions for persistence. Holling (1973) suggests that rather than concentrating on equilibrium states and worring about constancy and stability, we should shift our attention toward acquiring a better understanding of the domains of attraction and the boundaries of these domains (see the phase diagram of two species X and Y shown in Figure 3), and also concern ourselves more with the persistence of

---

1/ Stability is defined here as the ability of the ecosystem to return to an equilibrium state after a temporary disturbance. Resilience is defined as the ability of the ecosystem to absorb changes of state variables, driving variables, and parameters, and still persist. The domain of attraction is a region within which the ecosystem is stable and outside of which the system is unstable. See Holling (1973).

systems in the face of the unexpected, rare events--either natural or man-induced. Fiering and Holling (1974) have carried the idea a little further in suggesting that environmental standards should not be based solely on stability criteria using upper, or lower, bounds set in the political (social choice) process, but that natural systems should be allowed to swing freely over relatively wide extremes located with respect to the boundaries of the natural domains of attraction. This focuses the attention on persistence rather than on stability. They point out that some of the most persistent systems are relatively unstable, and vice versa.

Related to this, Fiering and Holling also raised an interesting economic issue: should we invest resources in "fail safe" engineering systems, or should we accept periodic failures and accidents that result in perhaps the complete collapse of a localized ecosystem. In the latter situation, the system would be reestablished from flora and fauna taken from an "environmental zoo". These issues and concepts are extremely relevant to the management of natural systems, and we hope that the speakers and discussants at this Symposium will address some of them.

Summary of paper

The purpose of this Symposium is to discuss the development of ecological models that will provide useful information for basing public policy regarding the management of our natural resources and to present some examples of ecological models that have been developed with this purpose in mind. In section II of this paper, we provided a conceptual

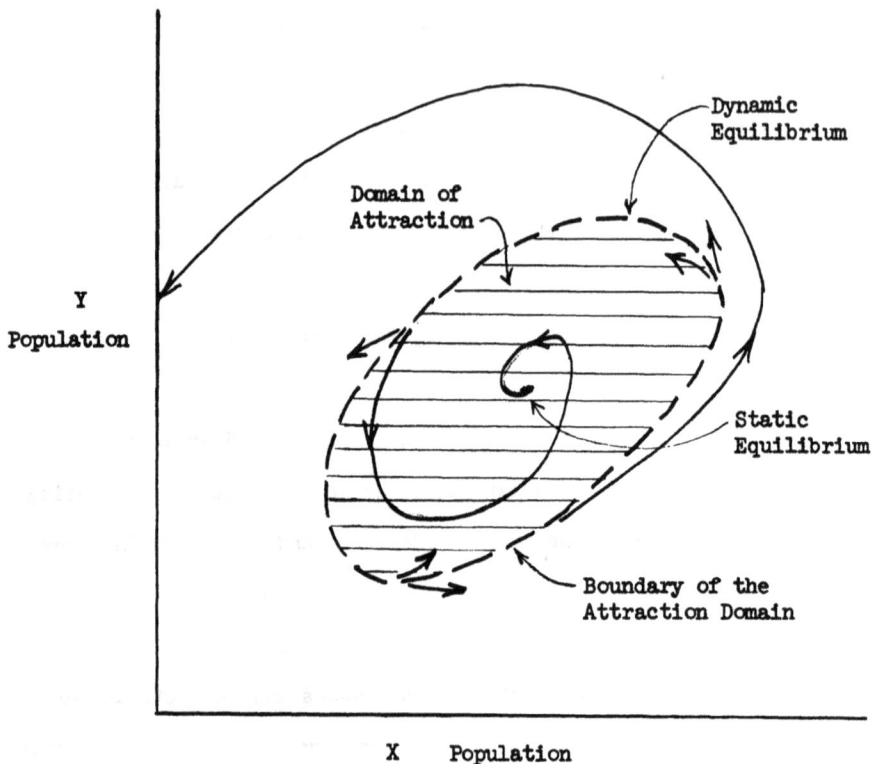

Figure 3. Phase Plane for Two Species X and Y Indicating Domain
of Attraction and Boundary of the Attraction Domain.

Source: Modification of Figure 2d in C. S. Holling, "Resilience and Stability of Ecological Systems," Annual
Review of Ecology and Systematics, Vol. 4 (1973),
pp. 1-23.

framework for analyzing regional environmental quality management problems and discussed some modeling approaches--both simulation and optimization--that could be used for ranking alternative management strategies. In section III, we concentrated on the use of mathematical programming (optimization) models for selecting "optimal" management strategies and indicated how nonlinear ecosystem models could be included within these management models using what we refer to as the "environmental response matrix". This discussion was based on deterministic analyses--of both economic and environmental systems--and employed the equilibrium approach to the analysis of ecosystems. This section did not address the problem of the rare, random events, or consider the resilience properties of the ecosystem. Persistence was assumed to prevail.

Questions for Symposium participants

There are two broad areas of research in ecosystem modeling and management that will occupy ecologists and water resources engineers for many years to come: (1) the development of ecological models in a resource management context that are able to provide society with useful information upon which to base public policy decisions regarding the uses, and levels of use, of our water resources, and (2) the development of ways of dealing with natural and man-induced uncertainties and surprises through management practices that will increase resilience and, hence, the probability that a given ecosystem will persist over time. With respect to these areas of research, though, it would be extremely helpful to us if we had a better idea of where we stood today. What is currently known, where are the gaps in our knowledge, and where will we be in ten, or fifteen, years from now? To assist us in assessing the

current state-of-the-art in ecosystem modeling, the following questions
are asked of the Symposium participants, especially of the speakers and
the discussants.

State-of-the-art and prospects for further advances

1. How well do the current models predict the state of the natural world under different management scenarios and types of
   residuals discharges? How well do they predict dissolved oxygen levels, algal densities, and fish biomass? What are the
   prospects for further advances? How much better (however defined) could we make these models with additional resources--
   money, competent researchers, data, etc.? What would it take
   to make them better predicting devices? What are the current
   stumbling blocks: lack of data (on inputs, state variables,
   parameter values, etc.); or lack of knowledge regarding material transfers between components ("compartments") of the ecosystem and the functional representations thereof; or what?

2. What are the prospects for verifying (as opposed to calibra-
   ting) ecological models? Are sufficient data currently available?

3. How can we simplify ecological models but maintain the important (from a social point of view) characteristics of the system? What are the tradeoffs among spatial scale, temporal
   scale, number and types of state variables?

4. Are stochastic aspects (errors in measurement, random variations of nature, random man-induced events) important? Should
   they be considered? Under what circumstances? Is it premature

to include these random fluctuations in current ecosystem
models?

5. Where would we get the largest payoff in supporting the devel-
    opment of ecological models: more and better data; research
    on material transfer rates between ecosystem components ("com-
    partments"); or what?

Experiences of the modelers of this Symposium

1. What are the strengths and weaknesses of your models; what are
    their limitations?

2. What kinds of tradeoffs were considered in developing your
    models?

3. What kinds of problems did you face in developing them?

4. Were you limited by data availability; available resources
    (money and trained personnel); current knowledge about physi-
    cal, chemical, and biological processes; or numerical solution
    problems?

A final question

Do you think we should continue to concentrate only on equilibrium
conditions and states (dynamic as well as static) in ecological modeling,
or should we give more attention to the domains of attraction and the
boundaries of these domains; to management strategies that do not force
systems to function near these boundaries (e.g., requiring high DO but
with no consideration given to the maintenance of other aspects of the
system); and to the whole question of the persistence of systems in the
face of the unexpected, rare events--either natural or man-induced?
What are your reactions to maintaining an "environmental zoo" for rees-
tablishing the flora and fauna of a collapsed ecosystem?

References

Bloom, S. G., A. A. Levin, and G. E. Raines. "Mathematical Simulation of Ecosystems: A Preliminary Model Applied to a Lotic Freshwater Environment," Battelle Memorial Institute, Columbus, Ohio, 16 April 1969.

Camp, Thomas R.. Water and Its Impurities (New York: Reinhold Publishing Co., 1963).

Chen, Carl W.. "Concepts and Utilities of Ecologic Model," Journal of the Sanitary Engineering Division, Proceedings of the American Society of Civil Engineers, SA 5 (Oct. 1970), pp. 1085-1097.

Custer, S. W., and R. G. Krutchkoff. Stochastic Models for Biochemical Oxygen Demand and Dissolved Oxygen in Estuaries, Bulletin 22, Water Resources Research Center, Virginia Polytechnic Institute, Blacksburg, Virginia, February 1969.

DiToro, Dominic M., Donald J. O'Connor, and Robert V. Thomann. "A Dynamic Model of Phytoplankton Populations in Natural Waters," Environmental Engineering and Sciences Program, Manhattan College, Bronx, New York, June 1970.

Dobbins, W. E. "BOD and Oxygen Relationships in Streams," Journal of the Sanitary Engineering Division, Proceedings of the American Society of Civil Engineers, SA 3 (June 1964), pp. 53-79.

Fiacco, Anthony V., and Garth P. McCormick. Nonlinear Programming: Sequential Unconstrained Minimization Techniques (New York: John Wiley & Sons, Inc., 1968).

Fiering, Myron B., and C. S. Holling. "Management and Standards for Perturbed Ecosystems," Working Paper No. 23, International Institute for Applied Systems Analysis, Laxenburg, Austria, February 1974 (to appear in the IIASA Research Report Series).

Haefele, Edwin T. Representative Government and Environmental Management (Baltimore: The Johns Hopkins University Press, 1973).

_____, ed. The Governance of Common Property Resources (Baltimore: The Johns Hopkins University Press, 1974).

Holling, C. S. "Resilience and Stability of Ecological Systems," Annual Review of Ecology and Systematics, Vol. 4 (1973), pp. 1-23.

Jansson, Bengt-Owe. Ecosystem Approach to the Baltic Problem, Bulletin No. 16 from the Ecological Research Committee of the Swedish Natural Science Research Council (NFR), Stockholm, Sweden, 1972.

Kelly, Robert A. "Conceptual Ecological Model of the Delaware Estuary," to be published in Bernard C. Patten, ed., Systems Analysis and Simulation in Ecology, Vol. IV (New York: Academic Press, forthcoming).

_____, and Walter O. Spofford, Jr. "Application of an Ecosystem Model to Water Quality Management: The Delaware Estuary," to be published in Charles A. S. Hall and John W. Day, eds., Models as Ecological Tools: Theory and Case Histories (New York: Wiley-Interscience, forthcoming).

Manhattan College. "Nitrification in Natural Water Systems," Environmental Engineering and Sciences Program Technical Report, Bronx, New York, 1970.

McCormick, Garth P. "Penalty Function Versus Non-Penalty Function Methods for Constrained Nonlinear Programming Problems," Mathematical Programming, Vol. 1, No. 2 (November 1971), pp. 217-238 (Amsterdam: North-Holland Publishing Company).

Nixon, Scott W., and James N. Kremer. "Narragansett Bay: The Development of a Composite Simulation Model for a New England Estuary," to be published in Charles A. S. Hall and John W. Day, eds., Models as Ecological Tools: Theory and Case Histories (New York: Wiley-Interscience, forthcoming).

O'Connor, Donald J. "The Temporal and Spatial Distribution of Dissolved Oxygen in Streams," Water Resources Research, Vol. 3, No. 1 (1st Quarter 1967), pp. 65-79.

Parker, Richard A., "Simulation of an Aquatic Ecosystem," Biometrics, Vol. 24, No. 4 (December 1968), pp. 803-821.

Patten, Bernard C. "Mathematical Models of Plankton Production," Internationale Revue Gesamten Hydrobiologie, Vol. 53, No. 3 (1968), pp. 357-408.

Russell, Clifford S. "Models for the Investigation of Industrial Response to Residuals Management Action," The Swedish Journal of Economics, Vol. 73, No. 1 (1971).

_____, and Allen V. Kneese. "Establishing the Scientific, Technical, and Economic Basis for Coastal Zone Management," Coastal Zone Management Journal, Vol. 1, No. 1 (1973), pp. 47-63.

_____, Walter O. Spofford, Jr., and Edwin T. Haefele. "The Management of the Quality of the Environment," in Jerome Rothenberg and Ian G. Heggie, eds., The Management of Water Quality and the Environment (New York: Halsted Press, 1974).

Sjöberg, S., F. Wulff, and P. Wåhlström. "The Use of Computer Simulations for Systems Ecological Studies in the Baltic," Ambio, Vol. 1, No. 6 (December 1972). (Stockholm: Royal Swedish Academy of Sciences)

Sobel, Matthew J. "Water Quality Improvement Programming Problems," Water Resources Research, Vol. 1, No. 4 (4th Quarter 1965), pp. 477-487.

Spofford, Walter O., Jr. "Total Environmental Quality Management Models," in Rolf A. Deininger, ed., Models for Environmental Pollution Control (Ann Arbor, Mich.: Ann Arbor Science Publishers, Inc., 1973).

_____, Clifford S. Russell, and Robert A. Kelly. "Operational Problems in Large-Scale Residuals Management Models," in Edwin S. Mills, ed., Economic Analysis of Environmental Problems (New York: National Bureau of Economic Research, 1975).

Streeter, H. W., and E. B. Phelps. "A Study of the Pollution and Natural Purification of the Ohio River," Public Health Bulletin No. 146, U.S. Public Health Service, Washington, D.C., 1925.

Thomann, Robert V., Donald J. O'Connor, and Dominic M. DiToro. "Modeling of Nitrogen and Algal Cycles in Estuaries," paper presented at 5th International Water Pollution Research Conference, San Francisco, California, 1970.

U.S. Environmental Protection Agency. Stochastic Modeling for Water Quality Management (Washington, D.C.: U.S. Government Printing Office, 1971).

CLEANER: The Lake George Model

Richard A. Park, Don Scavia, Nicholas L. Clesceri

ABSTRACT

CLEANER, an ecosystem model based on the International Biological Program model CLEAN, has a number of characteristics useful to environmental management. It represents functional physiologic and ecologic relationships for major compartments of the ecosystem, with disaggregation of trophic levels appropriate for studying competition among dissimilar forms. It exhibits good calibration and has few data requirements, facilitating transferability. It is programmed for use in interactive mode from remote terminals, with user-oriented output - including transformation of biomass values to turbidity, scum, and taste and odor indicators. It is currently implemented as a one-dimensional model without physical mixing terms, but it can be coupled with existing hydrodynamic models.

As a research tool CLEANER can be used to test hypotheses concerning complex ecosystem linkages and to guide data collection. As a management tool it can be used to provide scenarios and to extract bivariate relationships between pollutants and ecosystem effects. The model can be used by citizen groups as an educational tool, by advisory groups as a means of examining environmental trade-offs, and by regulatory agencies as a means of determining sensitivities and evaluating environmental impacts. CLEANER will eventually be coupled with adjunct models that predict nutrient loadings and tourist response, permitting simulation of long-range environmental, social and economic impacts.

INTRODUCTION

The ecosystem model CLEANER (Comprehensive Lake Ecosystem ANalyzer for Environmental Resources) has been developed by Rensselaer's Fresh Water Institute at Lake George in response to the need for a resource management model with ecologic realism. It is based on the model CLEAN, which was formulated by a multidisciplinary team in the Eastern Deciduous Forest Biome, U.S. International Biological Program (Park and others, 1974) and implemented by the Fresh Water Institute (Scavia and others, 1974).

CLEAN presently comprises 29 coupled ordinary differential equations representing the more important compartments of lake ecosystems (Figure 1). Several versions have been implemented, including CLEANER and WINGRA[2] which is specific to the Lake Wingra, Wisconsin, IBP site (MacCormick and others, 1972). CLEANER presently consists of the 14 equations in CLEAN specific to the pelagic (open-water) portions of lakes, especially Lake George, New York, with output designed particularly for environmental managers.

Although CLEAN was developed primarily as a scientific tool to study ecosystem dynamics, the implementation of CLEANER has been guided by careful consideration of the characteristics that would be most useful in the application of the model to problems of environmental management. The principal objective of this paper will be to describe these characteristics.

FUNCTIONALITY

Probably the single most important characteristic of CLEAN is its embodiment of our current understanding of functional relationships. Model development has attempted to represent the more important aspects of key ecologic and physiologic processes in the

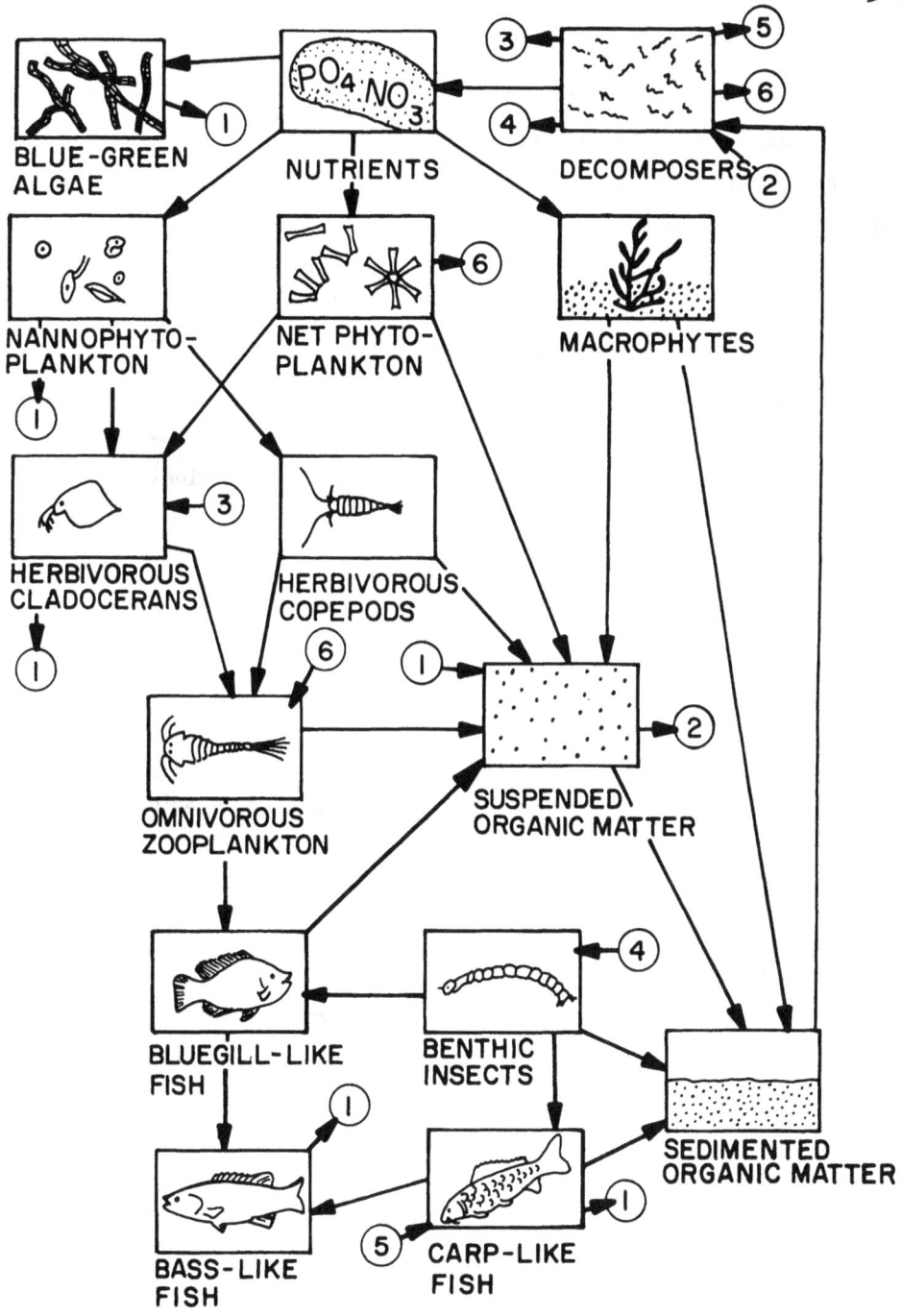

FIGURE 1. Ecosystem compartments represented in CLEAN.

mathematical formulation (Bloomfield and others, 1973; see also, Shugart and others, 1974). Of necessity these representations are greatly simplified, but they are based on a consideration of under-lying biologic principles. The result is a biomass model that exhibits a strong degree of ecologic realism and that is capable of being applied to a wide range of aquatic environments, with appre-ciable utility in environmental management.

Furthermore, the equations are formulated in such a way that they can be modified very easily. Each process is represented by a maximum-rate parameter (measured at optimal conditions) multi-plied by reduction factors for the effects of non-optimal conditions. For example, net photosynthesis is formulated as:

$$P_{net} = (P_{max} n/[\sum_i 1/\mu_i]-R)f(T)$$

P_{max} = maximum photosynthetic rate

n = number of limiting factors

μ_i = ith limiting factor: μ_1 = light, μ_2 = soluble nitrogen, μ_3 = phosphate

R = respiration rate

f(T) = complex function for effect of temperature

Rates can be changed by changing parameter values and by changing the formulation of the factors; additional factors can be incorporated simply by including them in the process equations. Thus, pesticide inhibition of photosynthesis could be represented as:

$$P_{net} = (P_{max} n/[\sum_i 1/\mu_i]-R)f(T)f(p)$$

f(p) = function for effect of pesticide with range from 0 (complete inhibition) to 1 (no inhibition)

SCOPE

Implicit in the development of CLEAN has been the realization that the whole ecosystem should be modeled, as also noted by Patten (1973). This has a direct bearing on the application of the model because environmental decision-making is increasingly being required to consider effects at all levels of the ecosystem. For example, the Federal Water Pollution Control Act (Public Law 92-500) calls for the restoration and maintenance of the biological integrity of the Nation's waters, with particular attention to the support of fish, shellfish, and wildlife and to the protection of public water supplies and waterbased recreational activities. If models of aquatic ecosystems are to be useful in investigating and forecasting the far-ranging consequences of pollution and physical disruption (including dredging, shoreline construction, and dumping) on such varied aspects of the ecosystem, they will have to represent not only algal growth, but the dynamics of zooplankton, fish, benthos, and bacteria as well.

The choice of compartments dictates the ultimate usefulness of the model. The use of too few compartments limits the way in which the model can represent the dynamic interactions found in the real world; the use of too many compartments results in undue requirements for parameter values and unnecessarily large computational loads. Very seldom would one want to simulate at the species level (Walters and Efford, 1972); but likewise, it is unrealistic to expect applicable results by modeling whole trophic levels, thereby reducing the food web to an overly simplified food "chain."

The choice of compartments modeled in CLEANER was based on a consideration of the principal modes of resource utilization and ecologic interaction found in the Lake George ecosystem. Therefore, the primary trophic level represented by phytoplankton was subdivided into three groups that reflect differences in nutrient utilization and susceptibility to grazing by zooplankton. In turn, zooplankton were divided into three groups on the basis of differences in feeding. The fish compartments also represent major feeding strategies, with additional compartments currently being added.

This degree of disaggregation permits environmental managers to observe the effects that man's impact will have on competition among ecologic groups, including both desirable and undesirable forms. For example, nutrient enrichment can favor the replacement of relatively innocuous nannophytoplankton by taste- and odor-producing large diatoms, which in turn may be replaced by scum-forming blue-green algae. The ascendency of these different types will affect the higher levels of the food web and can lead to a change in the fisheries - all of which can be represented to some degree by CLEANER. Likewise, the degree of disaggregation exhibited by CLEANER can permit the examination of the effects of alternate food pathways, such as plankton- and detritus-based systems, on the biological concentration and magnification of hazardous materials, including pesticides, heavy metals, and PCBs (Figure 2).

The equations are expressed in a general format (Park and others, 1974), so that if additional compartments are needed for a particular study they can be added with relative ease. Elaboration of the model is primarily a problem of parameterization.

In its present form CLEANER does not include a compartment for dissolved oxygen. Lake George is aerobic at all depths (Aulenbach, 1972); therefore, as a simplification, we have ignored the effects of dissolved oxygen. However, to ensure general applicability of the model, we are presently implementing dissolved oxygen as a state variable.

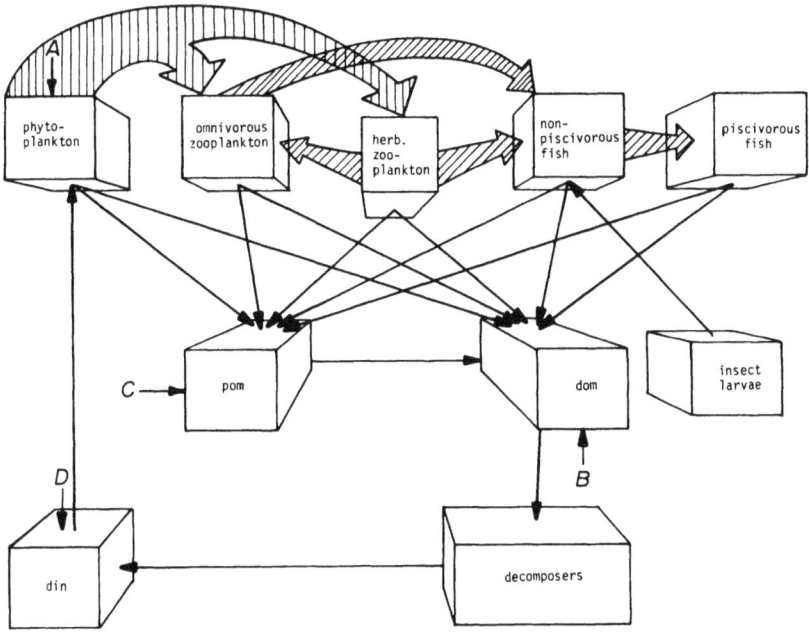

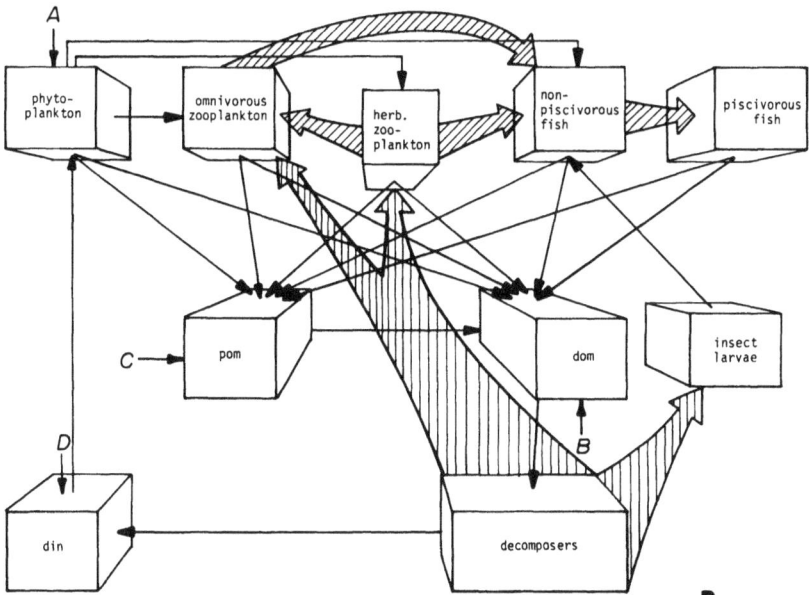

FIGURE 2. Flowchart of CLEANER showing: A. plankton-based food chain, and B. detritus-based food chain.

SPATIAL REPRESENTATION

CLEANER has been implemented as a one-dimensional model representing a m^2 column of water, which may be considered as indicative of average conditions for the lake or for a portion of the lake. The model differs from several popular ecosystem management models (for example, Chen, 1970; DiToro, O'Connor, and Thomann, 1971; Lombardo, 1971) in not including physical mixing terms, reflecting the emphasis on biologic processes in the IBP. Furthermore, as discussed below, it is felt that the resolution necessary for management of Lake George does not require continuous spatial modeling.

In the Lake George study multivariate analysis of the distributions of diatom frustules (siliceous skeletons) in bottom sediments has indicated the heterogeneity of nutrient enrichment of the lake (Bloomfield, 1972). As shown in Figure 3, the heterogeneity is relatively complex in the southern portion of the lake, but a basic pattern of enrichment adjoining the lakeside communities is evident. This information has suggested modelling the lake as homogeneous sub-basins. With this simplification, we can run the model for each major nutrient-enrichment segment of the lake, using sub-basin loadings and estimates of import and export values for nutrients. Furthermore, Lake George does not exhibit chemical stratification (Aulenbach, 1972); and renewal of the epilimnion by nutrient-rich hypolimnetic water during overturn is not a problem. Therefore, the effects of vertical mixing have been ignored.

On the other hand, there are management problems where spatial differences are of particular interest or where the complexities of shifting current patterns and stratification necessitate greater attention to the physical setting. Fortunately, physical modeling is relatively advanced. Therefore, one can either use existing

FIGURE 3. Nutrient-enrichment map of Lake George based on multivariate analysis of diatom death assemblages. After Bloomfield, 1972.

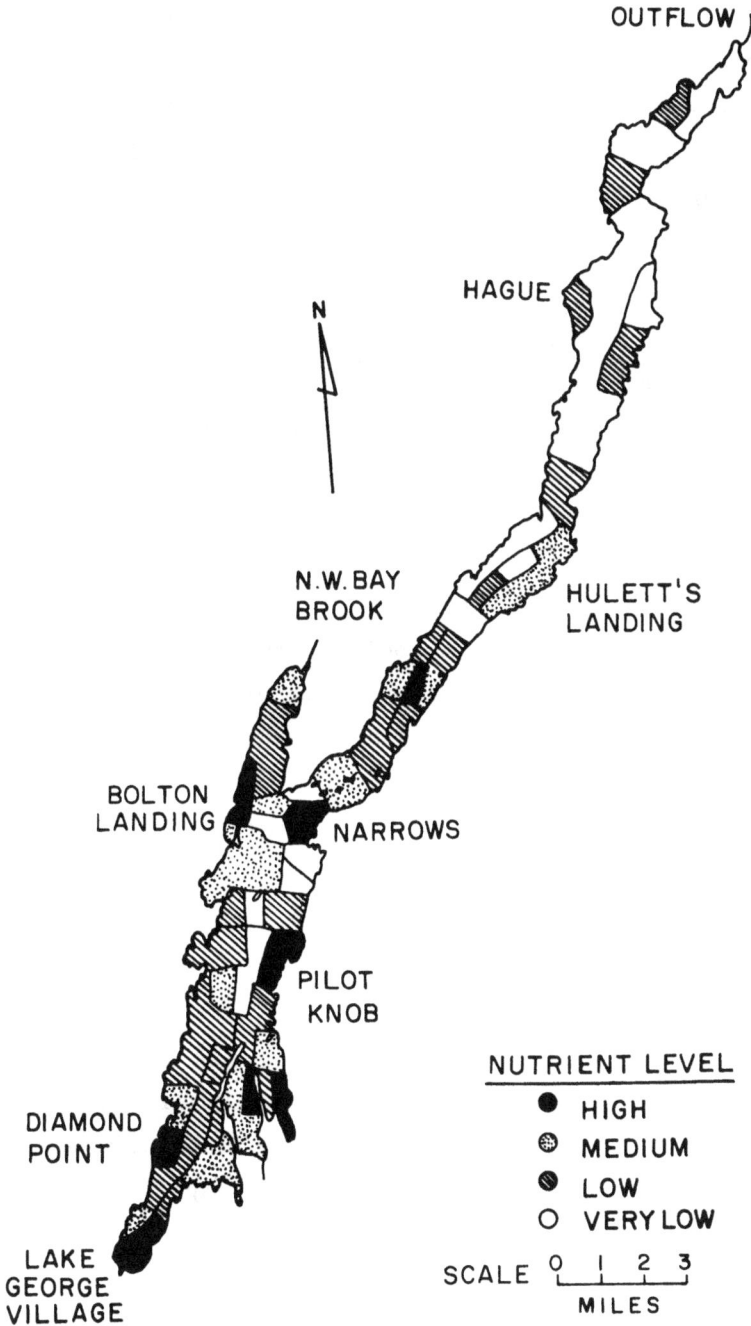

OUTFLOW

HAGUE

N

N.W. BAY
BROOK

HULETT'S
LANDING

BOLTON
LANDING

NARROWS

PILOT
KNOB

DIAMOND
POINT

LAKE
GEORGE
VILLAGE

NUTRIENT LEVEL

● HIGH
◉ MEDIUM
◍ LOW
○ VERY LOW

SCALE 0 1 2 3
MILES

hydrodynamic models to determine import-export rates from one spatial compartment to another through a given span of time for subsequent use in the ecosystem model, or one can couple the ecosystem model to a particular hydrodynamic model and run them simultaneously. For example, it is possible to use CLEANER in place of the biologic production term in the IBP transport model developed by Hoopes and others (In: Park and others, 1974).

CALIBRATION

The model has been calibrated using data from Lake George. The objective has been to approximate the seasonal patterns and mean levels of biomass for each of the compartments by varying parameter values within the ranges suggested by laboratory experimentation. No attempt was made to get an exact fit to the observations, especially in view of the sampling errors and stochastic variations inherent in the data.

The results are relatively good (Figure 4) with the exception of the decomposer simulation, which overestimates decomposer biomass. The difficulty here is probably one of parameterization: it does not seem to be feasible to obtain valid parameter estimates for all decomposer processes under controlled laboratory conditions. Additional field data will probably be required before this part of the model can be calibrated satisfactorily.

TRANSFERABILITY

Implementation of the model has emphasized those aspects that would insure its transferability. Although CLEANER has been implemented specifically for Lake George, it is based on a core of ecologically sound functions and is, therefore, expected to be

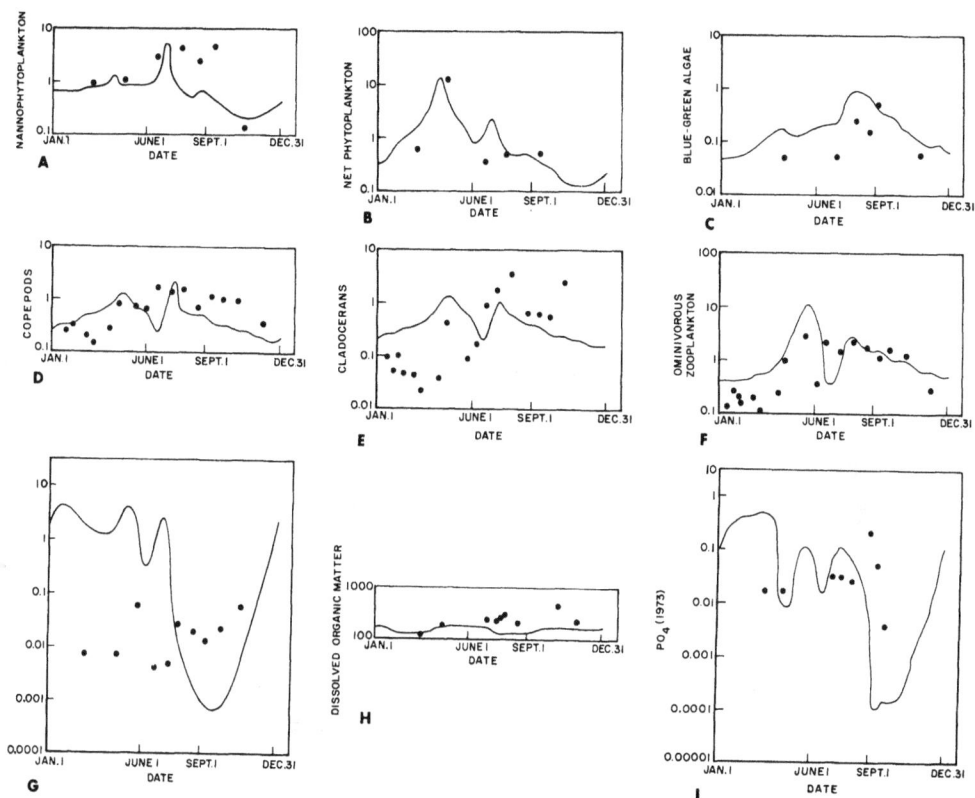

FIGURE 4. Comparison of results of simulation (solid lines) with data from Station 1, Lake George, New York (dots) for six compartments. Biomass and nutrients are in grams/m². Phytoplankton data from H. H. Howard, zooplankton data from D. C. McNaught, phosphate data from D. B. Aulenbach, dissolved organic matter data from S. Kobayashi, decomposer data from L. S. Clesceri and M. Dazé.

adaptable to other sites. During adaptation to a new site, the inclusion of hydrodynamic transport, vertical mixing, sediment resuspension and other mechanics should be included where necessary, as previously discussed.

Site Calibration - Since there are often differences in physiologic and ecologic responses at different sites, the model should be re-evaluated for each site. In addition to construct modification, it may be necessary to reparameterize the model to better represent endemic populations. We view this as a proper procedure where the present state of the lake is known and the objective is to forecast the results of perturbations.

By partitioning lake types that can be represented by specific model versions and particular sets of parameters, the fine-tuning phase can be reduced in many studies and the model can be applied even more readily. Some success has already been achieved with a related model using separate parameters for warm- and cold-water lakes (Walters, Park, and Koonce, in press).

Data Requirements - One of the difficulties in applying ecosystem models to various sites is that data are usually sparse. This was recognized early in the development of CLEANER and, accordingly, every effort has been made to reduce the amount of data required to run the model. At present, the driving variables include incident solar radiation (corrected for ice cover), water temperature (averaged for the water column), and loadings for nutrients and organic matter. Time-series for the first two can be approximated for a given site. The loadings can be measured directly or approximated using literature values (for example, Shannon and Brezonik, 1972). They can also be simulated using a separate model (see "Adjunct Models" below).

Obtaining reasonable initial-condition values for all compartments can be difficult. However, CLEAN has been shown to have good stability through successive years (Park and others, 1974); and, by using a "spin-up" period of three or more years, transient conditions caused by inappropriate initial conditions can be avoided. As shown in Table 1, the compartments in CLEANER will seek their proper levels even when the initial conditions are far from the actual values. Therefore, the only need for data on the individual ecosystem compartments is so that the model can be calibrated for a given site prior to perturbation analysis.

INTERACTIVE CAPABILITY

In order to maximize its accessibility and flexibility, CLEANER has been programmed for use on time-sharing systems from remote terminals. Thus, we are able to access the model from virtually any location. As a result, the model has been used to some advantage in seminars, workshops, and meetings at a variety of locations. Aside from its availability on one of our local computers, the model is being implemented on a commercial time-sharing system that is under contract to the Environmental Protection Agency, thus making it available to their personnel from anywhere in the country.

Equally important is the fact that users can interact with CLEANER by on-line editing of driving variables, site constants (such as water depth), initial conditions, and parameters. The time-series for nutrients, light, and temperature can be changed merely by setting the appropriate perturbation parameter. Consequently, entering a single statement:

PERT (1) = 2

on the terminal will double the phosphate loading for the period of

Table 1 - Initial and Final Biomass Levels After Eight Year Simulation
(g/m^2)

	INITIAL			FINAL		
	Run 1	Run 2	Run 3	Run 1	Run 2	Run 3
NANNOPLANKTON	.59	1.0	1.5	.18	.18	.18
NET PLANKTON	.34	1.0	1.5	.08	.08	.08
BLUE GREEN ALGAL	.05	.09	.002	.044	.047	.044
CLADOCERANS	.22	.1	.05	.11	.11	.11
COPEPODS	.26	.1	.05	.11	.12	.11
OMNIVORES	.37	.1	.05	1.03	.99	1.03
NON-PISCIVORES	.011	.3	.5	.02	.02	.02
PISCIVORES	.556	.3	.5	.767	.744	.767
PO_4	.07	1.0	.2	.005	.004	.005
NITROGEN	2.28	1.0	.2	2.65	2.63	2.65
POM	.22	.5	.2	.13	.13	.13
DOM	150.8	150.0	190.0	156.4	156.1	156.4
DEC	1.858	5.0	.8	.096	.069	.096

the simulation. Likewise, the user can explore the effects of a wide variety of perturbations such as halving the nitrogen input, raising the mean annual temperature by 10°C, removing all lake trout, and inhibiting photosynthesis in blue-green algae.

<u>User-oriented Output</u> - Considerable attention has been devoted to programming CLEANER so that the output would be meaningful to users other than the aquatic-specialist. The integration results are given in tabular form (Figure 5). Plots can be obtained in linear, semi-log, scaled or unscaled form showing any or all state variables (Figure 6).

It is not enough to create a model capable of mimicking these aquatic systems. If watershed managers and other non-specialist groups are to use the model, output must be in a more meaningful form. As optional output, therefore, a tabulation of key environmental-perception characteristics is available (Figure 7). These informative transformations of the biomass values include: predicted fish catches, transparency, and concentrations of scum-forming blue-green algae and taste- and odor-producing net phytoplankton, as well as algal ratios and concentrations of chlorophyll <u>a</u> with which environmental managers are accustomed to working.

APPLICATION

Although it is difficult to avoid using the word "predict," we prefer to think of CLEANER as being a "diagnostic" tool: one that is capable of diagnosing effects of man-induced perturbations. Ecosystems are very complex, characteristically with non-linear linkages among compartments; and it is not uncommon to find that perturbations produce counter-intuitive results. By using ecosystem models one can symbolically explore ecologic relationships and gain

```
♦INT
READY
DATE: 090374 (MMDDYY)  TIME: 095521 (HHMMSS) LAKE: GEO
```

TIME	NAN	NET	B-GRN	CLAD	COPE	OMNIZ	NON-P	PISC	PO4	NIT	POM	DOM	DEC
1	.59	.34	.050	.22	.26	.37	.011	.556	.070	2.28	.22	150.8	1.858
15	.63	.39	.045	.22	.26	.35	.010	.519	.201	3.73	.26	142.6	4.614
29	.68	.51	.043	.21	.27	.35	.009	.488	.356	5.39	.35	133.7	5.271
43	.76	.71	.042	.21	.27	.34	.008	.461	.457	6.56	.40	130.3	3.747
57	.85	1.04	.042	.21	.27	.33	.007	.437	.507	7.22	.36	131.5	2.423
71	.99	1.61	.043	.21	.27	.32	.007	.415	.524	7.51	.29	134.1	1.766
85	1.25	2.94	.044	.21	.27	.32	.006	.393	.500	7.36	.28	136.3	1.527
99	2.08	7.77	.046	.24	.29	.34	.005	.368	.335	5.55	.53	141.5	1.736
113	3.46	16.04	.045	.38	.40	.50	.005	.340	.024	1.74	1.58	150.1	3.625
127	4.00	18.17	.039	.94	.87	1.34	.005	.312	.022	.60	3.54	139.9	9.437
141	1.77	7.97	.043	2.04	1.78	6.16	.016	.283	.114	.20	9.10	126.7	3.930

```
----------------------------------MAX TIME. STEP=  2.6250 MAX ERR=   .06%
```

TIME	NAN	NET	B-GRN	CLAD	COPE	OMNIZ	NON-P	PISC	PO4	NIT	POM	DOM	DEC
155	.79	1.66	.048	.98	.91	9.47	.045	.254	.156	.38	4.08	128.6	.706
169	.73	.75	.052	.56	.55	9.35	.112	.226	.158	.45	1.19	131.6	.182
183	.78	.62	.056	.49	.49	8.21	.264	.202	.138	.42	.57	134.4	.058
197	.78	.57	.059	.43	.43	6.62	.559	.187	.110	.30	.51	136.3	.022
211	.76	.54	.058	.36	.37	4.74	.964	.184	.086	.22	.54	137.3	.010
225	.78	.51	.055	.28	.29	3.01	1.281	.193	.066	.19	.45	138.9	.005
239	.85	.50	.048	.22	.23	1.81	1.366	.208	.047	.20	.31	141.3	.003
253	.95	.47	.039	.17	.19	1.11	1.266	.222	.026	.20	.22	143.3	.002
267	.99	.42	.031	.14	.15	.72	1.087	.228	.010	.22	.16	144.9	.003
281	.89	.35	.023	.11	.13	.50	.899	.227	.003	.30	.12	146.0	.004
295	.73	.27	.017	.10	.11	.37	.732	.218	.002	.44	.13	147.3	.012
309	.65	.22	.014	.08	.10	.29	.597	.208	.004	.64	.12	149.4	.053
323	.70	.20	.011	.07	.09	.23	.492	.196	.021	1.00	.09	151.2	.354
337	1.00	.25	.009	.07	.08	.20	.411	.184	.118	2.04	.11	146.8	2.282
351	1.52	.42	.009	.07	.08	.19	.348	.174	.358	4.19	.14	131.4	5.046
365	1.82	.61	.010	.07	.09	.18	.300	.165	.524	5.78	.17	123.2	3.258

```
READY
```

FIGURE 5. Tabular output of integration results from CLEANER.

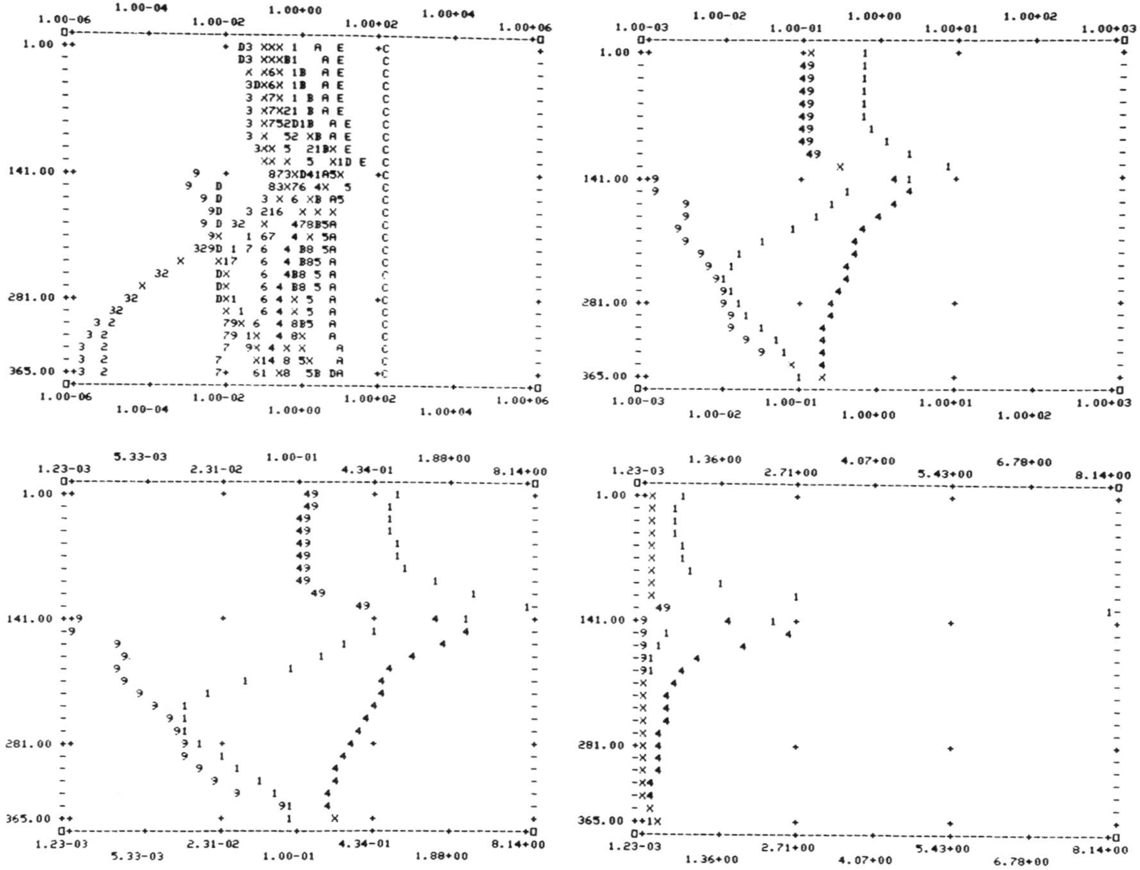

FIGURE 6. Plot output from CLEANER, demonstrating various options for scaling and for suppressing state variables. Simulation is for one year (vertical axis); biomass values are given on the horizontal axes.

◆TAB;A
ENVIRONMENTAL PERCEPTION PARAMETERS;   LAKE CODE=GEO   AVG DEPTH=30.0

                2-WEEK FISH CATCH
TIME N-PISC PISC SECCHI DISC ALGAE NET/NAN BLGRN/OTHER NET BLGRN CHLA
        #   #   METERS   MG/L   MG/L MG/L MG/L
----
   1.   9.   0. 40.31   .07   .576   .054   .005 .050 .001
  15.   8.   0. 46.92   .06   .629   .044   .005 .045 .001
  29.   7.   0. 46.00   .06   .748   .036   .007 .043 .001
  43.   7.   0. 47.49   .06   .939   .029   .010 .042 .001
  57.   6.   0. 48.65   .07 1.214   .022   .014 .042 .001
  71.   5.   0. 48.67   .08 1.628   .016   .022 .043 .001
  85.   5.   0. 46.94   .10 2.353   .010   .041 .044 .001
  99.   4.   37. 39.39   .21 3.733   .005   .128 .046 .002
 113.   4.   34. 24.31   .57 4.638   .002   .429 .045 .005
 127.   4.   32.   9.72 1.52 4.548   .002   1.218 .039 .013
 141.   13.   29. 15.60   .45 4.517   .004   .333 .043 .004
 155.   37.   26. 32.55   .10 2.112   .019   .033 .048 .001
 169.   92.   23. 38.58   .08 1.036   .035   .013 .052 .001
 183. 217.   20. 40.74   .08   .793   .040   .010 .056 .001
 197. 461.   19. 42.67   .08   .729   .043   .009 .059 .001
 211. 794.   19. 45.11   .08   .704   .045   .008 .058 .001
 225. 1055.   19. 47.75   .07   .661   .042   .007 .055 .001
 239. 1126.   21. 49.99   .07   .585   .036   .006 .048 .001
 253. 1043.   22. 51.76   .06   .495   .028   .006 .039 .000
 267. 895.   23. 53.27   .05   .426   .022   .005 .031 .000
 281. 740.   23. 54.64   .04   .391   .019   .004 .023 .000
 295. 603.   22. 55.71   .03   .373   .017   .003 .017 .000
 309. 492.   0. 56.41   .02   .341   .016   .003 .014 .000
 323. 405.   0. 56.42   .02   .291   .012   .002 .011 .000
 337. 338.   0. 53.70   .02   .248   .007   .003 .009 .000
 351. 287.   0. 49.26   .03   .276   .005   .006 .009 .000
 365. 247.   0. 50.85   .04   .335   .004   .008 .010 .000

TOTAL NON-PISC CATCH FROM THE BASIN IS   8909.
TOTAL PISCIVOROUS CATCH FROM THE BASIN IS   369.
READY

FIGURE 7.   Tabular output of environmental-perception
            characteristics from CLEANER.

a feeling for the way the system probably would react to a particular perturbation.

As a research tool, CLEAN can be used to test hypotheses concerning the ecosystem linkages that are not otherwise amenable to experimentation. It can also guide future research and data collection by explicitly defining those areas needing further description (Park and others, 1974). The history of development of CLEAN amply demonstrates this (Park and Wilkinson, 1971a, 1971b; Park and others, 1972, 1973; Bloomfield and others, 1973).

After several years of development, CLEAN, as manifested in CLEANER is reaching the stage where it can be used as a management tool as well as a research and educational tool. It does not provide answers, but it does provide a wealth of information on the effects of alternative management schemes, permitting interested groups to make their own value judgments. In effect, the model provides scenarios that can supplant the very slow feedback that characterizes the actual management and monitoring of the ecosystem. Furthermore, by varying the intensity of perturbation one can gain an understanding of the sensitivity of the model - and, by extension, of the real world - in that particular factor-space.

As an example, consider the use of the model in studying the effects of phosphorous enrichment. Various intensities of phosphorous loadings were used in a series of simulations of the south end of Lake George. Maximum phytoplankton biomass and minimum transparency were plotted against phosphorous loads (Figure 8). It can be seen that, in the extreme case, blue-green algae completely dominated the system; and even in less extreme situations the most dramatic change is in the ascendency of those noxious algae. Also, the water quality decreased markedly with increasing phosphorous loadings, as indicated by the simulated secchi disc readings. In

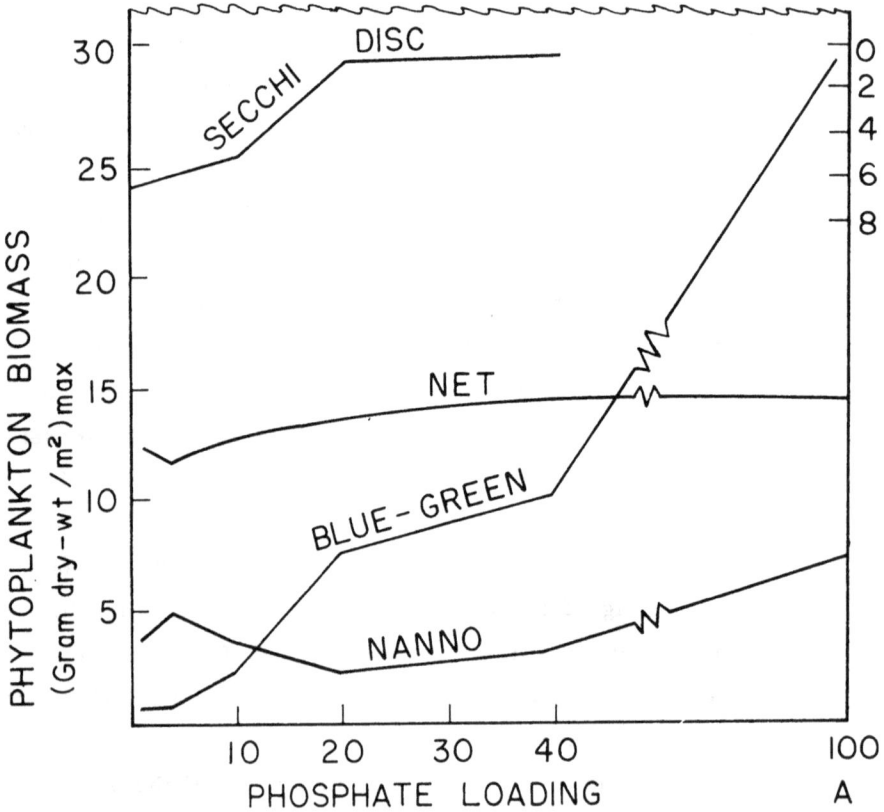

FIGURE 8. Maximum predicted biomass of net phytoplankton,
nannophytoplankton, and blue-green algae and
minimum predicted secchi disc readings plotted
against phosphate loadings (multiplicative with
normal phosphate loading from the drainage
basin of the southern portion of Lake George.)

this way, the model can be used to establish simple relationships between algal problems and phosphorous loadings that can be readily applied by environmental policymakers and managers. It is interesting to note that when excess nitrogen loadings were simulated, there was little response (Figure 9), as would be expected in that Lake George has been shown to be primarily phosphorous-limited for much of the growing season (Stross, 1971).

We anticipate that CLEANER will be used somewhat differently by various elements of the environmental management system (Figure 10). For instance, we have found that interested citizens tend to respond to the model output according to their particular interests, with fishermen looking for increased productivity and cottage owners worrying about water clarity and taste (especially on Lake George where most drinking water is pumped directly from the lake). For these special-interest groups the model can be used as an educational tool, giving them more objective insights into the intricacies of the aquatic ecosystem and their relationships to it.

Similarly, CLEANER is appropriate for legislative advisory groups that are concerned with the trade-offs involved in managing the environment. The model should facilitate the formulation of goals by providing scenarios based on alternative policies. Advisory groups seem to be particularly interested in cost-benefit relationships; therefore, linkages of a socio-economic model to the ecosystem model, as discussed below under "Adjunct Models," would be invaluable.

Theoretically, CLEANER and similar models should be quite useful to environmental managers. At the present time the implementation of regulations seems to be manifested primarily in the review of environmental standards and the evaluation of effects; it would appear that the model would be most helpful in differentiating

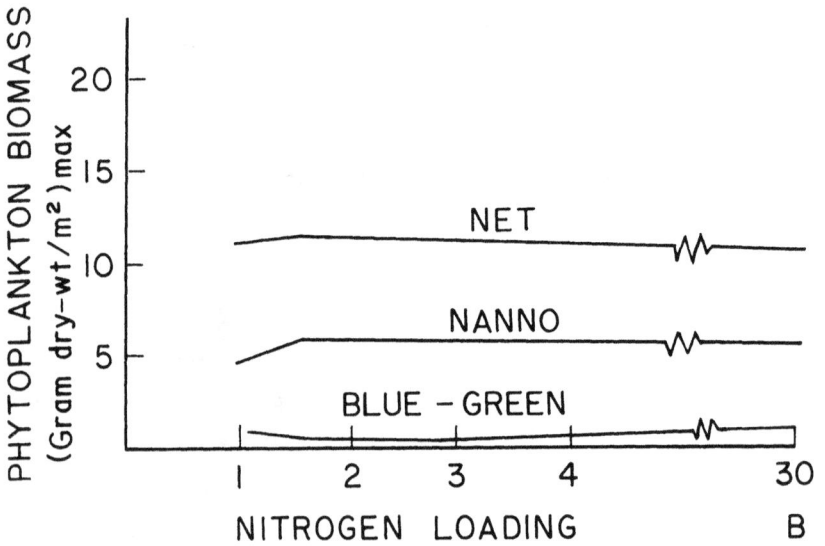

FIGURE 9. Same as in Figure 8, except that nitrogen loading was varied instead of phosphate.

71

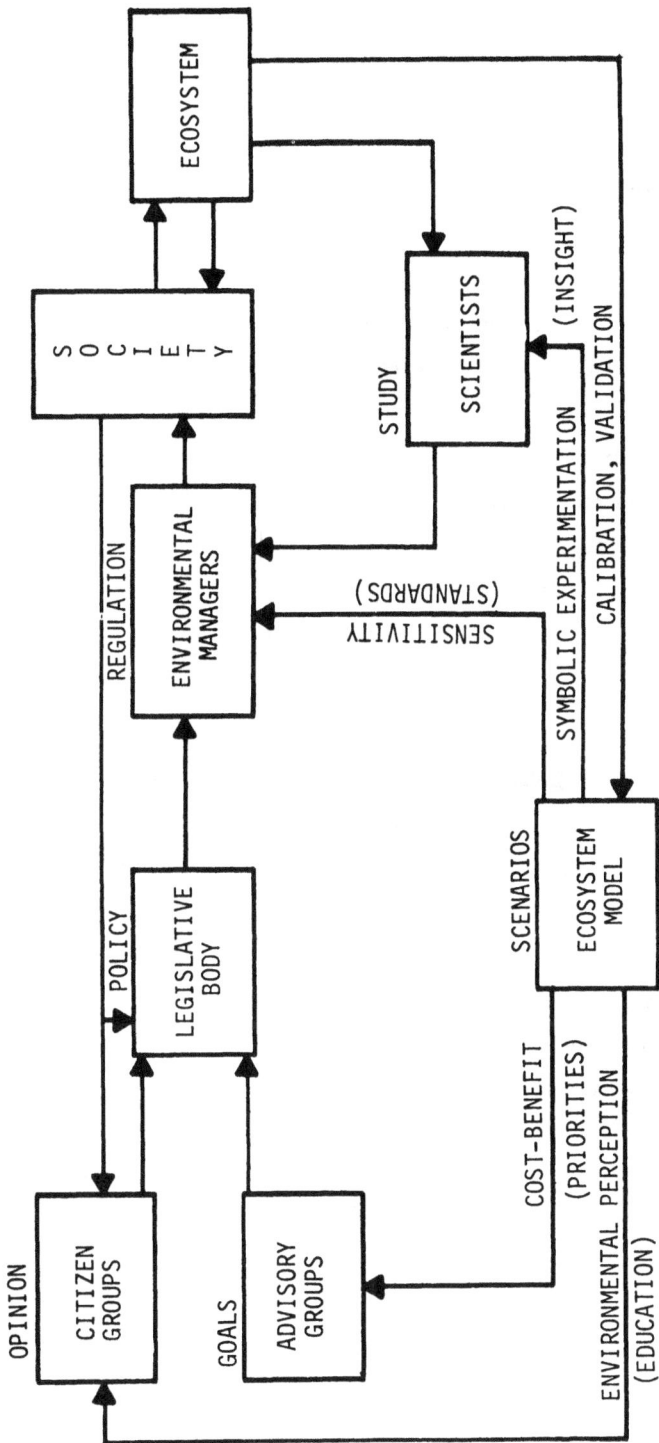

FIGURE 10. Relationship of elements of the environmental management system to CLEANER.

between sensitive and insensitive ranges of pollutants, based on whole-ecosystem response. The perturbation analysis described above would be useful in determining these sensitive ranges and critical points. In the same fashion, CLEANER can be used either directly or indirectly in the evaluation of environmental impact statements (EIS). The effects of perturbations on the system can be investigated and results can be expressed as simple bivariate relationships, whereby a relatively expedient assessment of the EIS can be made. In writing an EIS, where more time and manpower is often available, the model can provide an in-depth diagnosis of the ecosystem responses to the anticipated stresses. In this way the effects throughout the food web can be examined, and a thorough assessment of the impact can be delineated.

ADJUNCT MODELS

The usefulness of CLEANER as an environmental management tool will be enhanced when it is coupled with adjunct models developed in the course of the Lake George study (Park and Wilkinson, 1971a; Clesceri and Ferris, 1971). The concept of the linkages among these models might well serve as an example for modeling programs in general.

In order to examine the effects that changing land-use patterns and construction of waste-treatment facilities will have on nutrient input in a given basin, we are implementing WTRSHD, a nutrient/ hydrology model (Holberger, personal communication). The model considers both point and non-point sources of nutrients and the transport of these nutrients as a function of surface and groundwater flow (Figure 11). The final version of WTRSHD will obviate the need for data on nutrient loadings; it will require time-series data on precipitation and populations of permanent residents and

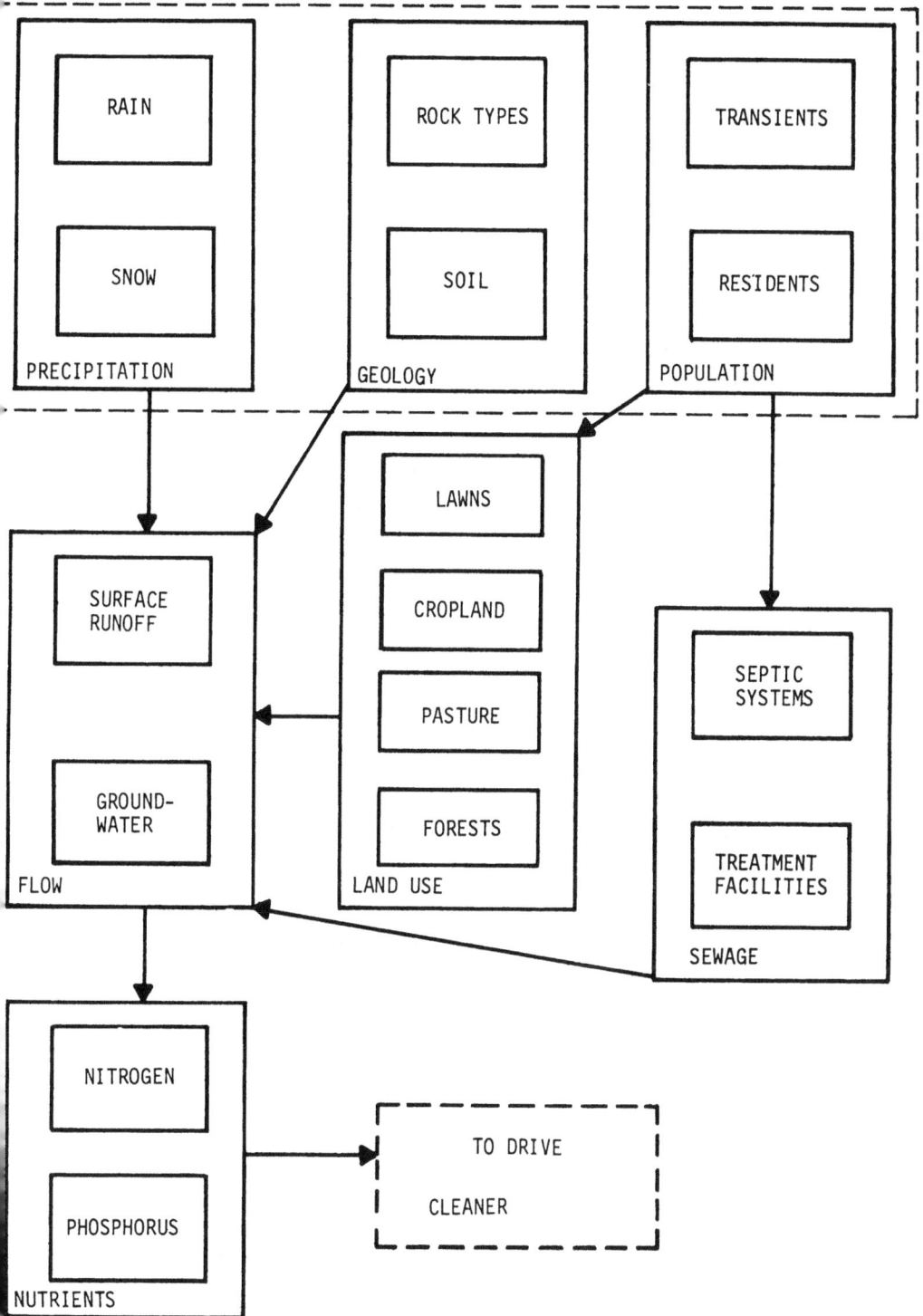

FIGURE 11. Major components of the nutrient/hydrology model WATRSHD.

transients and data on areal extent of major rock and soil types, and land usage - all of which can be obtained relatively easily for most lake basins.

A second model, POPUL, provides a means for predicting the growth of populations of transients and permanent residents and the resulting changes in land-use patterns (Figure 12). At present, the model calculates nutrient loadings and represents negative feedback to transient population by means of overly simplified formulations (Stern, 1971). However, POPUL can provide the land-use and population input for WTRSHD which, in turn, can calculate the nutrient input to CLEANER using well established functionalities.

Many people have expressed concern regarding the quantification of the intangible aspects of water-quality aesthetics. Because little is known about this important subject, environmental perception by recreationists, cottage and homeowners, and businessmen has been extensively studied at Lake George and three other lakes with dissimilar characteristics (Kooyoomjian and Clesceri, 1974). The recreationists were subdivided into boating, camping, fishing, picnicking, swimming, sightseeing, and amusement-park categories. Data are available showing how each of these groups perceive numerous aspects of the lake environment; considering that the response may be positive or negative, the data can be used to predict differential usage patterns for a range of water-quality states. Therefore, response to water quality can be modeled with some assurance. With prediction of water quality by CLEANER, the feedback loop to POPUL is formed and the linkage of the three models is complete (Figure 13).

Furthermore, the perception data show spending patterns by categories and water-quality states; these will eventually be used in developing an economic model to predict the economic consequences

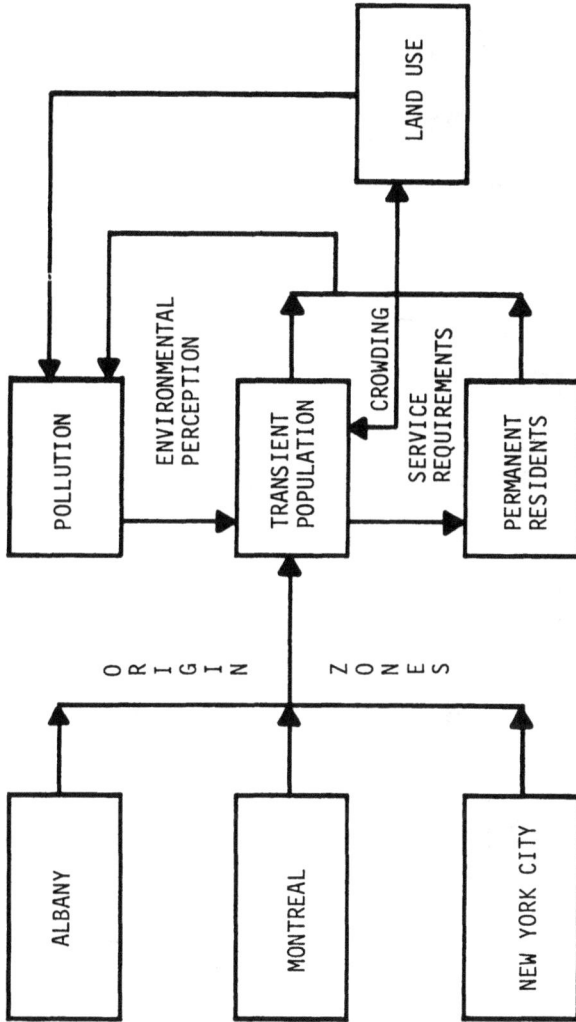

FIGURE 12. Major components of the population model POPUL.

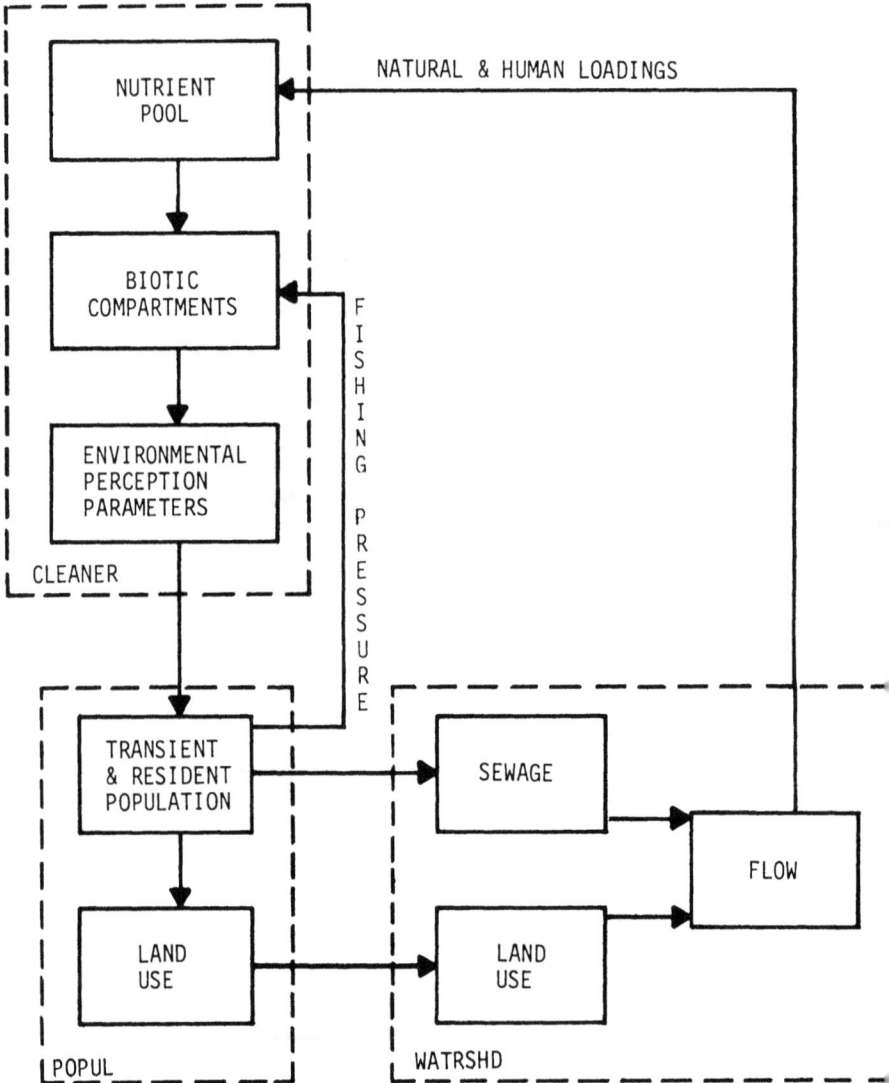

FIGURE 13. Conceptual linkage of the models CLEANER,
POPUL and WATRSHD.

of changing water qualities.   For example, we know that recreationists, other than fishermen, at oligotrophic (nutrient-poor) lakes
spend more per year than recreationists at eutrophic (nutrient-rich)
lakes (Kooyoomjian, 1974); thus the "cost" of degrading water quality
can be assessed.

By developing this family of compatible models, a sophisticated,
objective management tool can be created to examine the long-range
environmental, social, and economic impacts at recreational lake
sites.

SUMMARY

CLEANER exemplifies several characteristics that are useful
in applied modeling efforts.   These include: 1) functionality, em-
bodying the more important aspects of key ecologic and physiologic
processes in a formulation that can be modified easily; 2) broad
scope, including simulation of the major ecosystem compartments,
with enough disaggregation to permit observation of competition
between desirable and undesirable forms and examination of the
effects of concentration of hazardous substances through alternate
food pathways; 3) transferability, with few data requirements and
with a formulation that facilitates reparameterization; 4) interactive
capability, permitting on-line editing of driving variables, site
constants, initial conditions, and parameters from remote terminals;
and 5) user-oriented output, including a variety of plotting options,
and tabulations of biomass values and of environmental-perception
characteristics.

The model is a useful tool for scientific research, for educa-
tion, and for environmental management.   It is capable of providing
detailed scenarios for numerous types of perturbations and can help
in 1) diagnosing basic ecologic relationships, 2) providing a learning

experience for concerned citizenry, 3) examining environmental
trade-offs, 4) determining critical values of pollutants, and 5) evaluating environmental impacts of man-induced stresses.

The modeling program has also led to the development of
adjunct models to simulate hydrology and nutrient loadings and citizen response; and data are available for development of an economic
model. These will eventually be coupled to CLEANER in order to
simulate long-range environmental and socio-economic impacts.

ACKNOWLEDGMENTS

We wish to thank all those individuals who contributed to the
philosophy, formulation, and implementation of CLEANER and its
antecedent, CLEAN, especially R. V. O'Neill, J. A. Bloomfield,
J. S. Fisher, C. S. Zahorcak, and J. W. Wilkinson. The help of
J. J. Ferris in critically reviewing the manuscript is also acknowledged.

Research supported in part by the Eastern Deciduous Forest
Biome, U.S. International Biological Program funded by the National
Science Foundation under Interagency Agreement AG 199, 40-193-69
with the Atomic Energy Commission, Oak Ridge National Laboratory.

BIBLIOGRAPHY

Aulenbach, D. B., 1972, Chemical Nutrients in Lake George: U.S.
    International Biological Program, Eastern Deciduous Forest
    Biome Memo Report 72-63, 69 pp.

Bloomfield, J. A., 1972, Diatom Death Assemblages as Indicators of
    Environmental Quality in Lake George, New York: unpubl.
    master's thesis, Rensselaer Polytechnic Institute, Troy, N.Y.

Bloomfield, J. A., R. A. Park, Don Scavia, and C. S. Zahorcak,
    1973, Aquatic Modeling in the Eastern Deciduous Forest Biome,

U.S. International Biological Program, In: E. J. Middlebrooks, D. H. Falkenborg, and T. E. Maloney (eds.): Modeling the Eutrophication Process, Utah State Univ., p. 139-158.

Chen, C. W., 1970, Concepts and Utilities of Ecologic Model: Jour. Sanitary Engineering Division American Society Chemical Engineering, Vol. 96, p. 1085-1097.

Clesceri, N. L., and J. J. Ferris, 1971, The Progress of the Lake George Water Research Center: U.S. International Biological Program, Eastern Deciduous Forest Biome Memo Rept. 71-20, 95 pp.

DiToro, D. M., D. J. O'Connor, and R. V. Thomann, 1971, A Dynamic Model of the Phytoplankton Population in the Sacramento-San Joaquin Delta, In: J. D. Hem (ed.), Nonequilibrium Systems in Natural Water Chemistry: American Chemistry Society, Advances in Chemistry Series, Vol. 106, p. 3-180.

Kooyoomjian, K. J., 1974, The Development and Implementation of a Questionnaire Survey Data Base for Characterizing Man-Environment Relationships in Trophically Polarized Fresh Water Recreational Environments: unpubl. doctoral dissertation, Rensselaer Polytechnic Institute, 875 pp.

Kooyoomjian, K. J., and N. L. Clesceri, 1974, Perception of Water Quality by Select Respondent Groupings in Inland Water-Based Recreational Environments: Water Resources Bulletin, American Water Resources Assoc., Vol. 10, p. 728-744.

Lombardo, P. S., 1971, A Mathematical Model of Water Quality in an Impoundment: unpubl. master's thesis, University of Washington.

MacCormick, A. J. A., O. L. Loucks, J. F. Koonce, J. F. Kitchell, and P. R. Weiler, 1972, An Ecosystem Model for the Pelagic Zone of Lake Wingra: U.S. International Biological Program, Eastern Deciduous Forest Biome Memo Report 72-122,

103 pp.

Park, R. A., R. V. O'Neill, J. A. Bloomfield, H. H. Shugart, Jr., R. S. Booth, R. A. Goldstein, J. B. Mankin, J. F. Koonce, Don Scavia, M. S. Adams, L. S. Clesceri, E. M. Colon, E. H. Dettmann, J. A. Hoopes, D. D. Huff, Sanuel Katz, J. F. Kitchell, R. C. Kohberger, E. J. LaRow, D. C. McNaught, J. L. Peterson, J. E. Titus, P. R. Weiler, J. W. Wilkinson, and C. S. Zahorcak, 1974, A Generalized Model for Simulating Lake Ecosystems: Simulation.

Park, R. A., and J. W. Wilkinson, 1971a, Lake George Modeling Philosophy: U.S. International Biological Program, Eastern Deciduous Forest Biome Memo Report 71-19, 62 pp.

Park, R. A., and J. W. Wilkinson, 1971b, Lake George Modeling Project: U.S. International Biological Program, Eastern Deciduous Forest Biome Memo Report 71-117, 60 pp.

Park, R. A., J. W. Wilkinson, J. A. Bloomfield, R. C. Kohberger, and C. L. Sterling, 1972, Aquatic Modeling, Data Analysis, and Data Management at Lake George, New York: U.S. International Biological Program, Eastern Deciduous Forest Biome Memo Report 72-70, 32 pp.

Park, R. A., J. W. Wilkinson, R. C. Kohberger, J. A. Bloomfield, C. S. Zahorcak, and Don Scavia, 1973, Statistical Analysis, Data Management and Ecosystem Modeling at Lake George, New York: U.S. International Biological Program, Eastern Deciduous Forest Biome Memo Report 73-64, 6 pp.

Patten, B. C., 1973, Need for an Ecosystem Perspective in Eutrophication Modeling, In: E. J. Middlebrooks, D. H. Falkenborg, and T. E. Maloney (eds.): Modeling the Eutrophication Process, Utah State Univ., p. 83-87.

Scavia, Don, J. A. Bloomfield, J. S. Fisher, James Nagy, and

R. A. Park, 1974, Documentation of CLEANX, A Generalized Model for Simulating the Open-Water Ecosystems of Lakes: Simulation.

Shannon, E. E., and P. L. Brezonik, 1972, Relationships Between Lake Trophic State and Nitrogen and Phosphorus Loading Rates: Environmental Science and Technology, Vol. 6, p. 719-725.

Shugart, H. H., Jr., R. A. Goldstein, R. V. O'Neill, J. B. Mankin, 1974, TEEM, A Terrestrial Ecosystem Model for Forests: Oecologia Plantarum.

Stern, H. I., 1971, A Model for Population-Recreational Quality Interactions of a Fresh Water Site: Rensselaer Polytechnic Institute, Operations Research and Statistics Research Paper 37-71-P4, 24 pp.

Stross, R. G., 1971, Primary Productivity in Lake George, New York: Its Estimation and Regulation: U.S. International Biological Program, Eastern Deciduous Forest Biome Memo Report 71-115, 46 pp.

Walters, C. J., and I. E. Efford, 1972, Systems Analysis in the Marion Lake IBP Project: Oecologia, Vol. 11, p. 33-44.

Walters, C. J., R. A. Park, and J. F. Koonce, in press, Systems Modelling, In: E. P. LeCren (ed.), Synthesis of IBP Freshwater Studies, Oxford Press.

A DISCUSSION OF

CLEAN, THE AQUATIC MODEL OF THE

EASTERN DECIDUOUS FOREST BIOME

By

Carl W. Chen
Tetra Tech, Inc.
Lafayette, California   94549

INTRODUCTION

The paper described a mathematical model CLEAN, a computer code name for the Comprehensive Lake Ecosystem Analyzer.
Detailed mathematical formulations of the model were presented
elsewhere [Bloomfield, et al, 1973, Park, et al, 1974].   In the
paper, general discussions were made for the model's functionality,
parameters included, spatial representation, transferability, capability for input perturbation, and output.   The paper also mentioned
how the model could be used in conjunction with a watershed model
for the performace of population and waste load computations.

The model was claimed useful both as a tool for ecosystem
research and as an evaluation methodology for environmental management of a lake.   The capability for interaction between users
and the computer model were emphasized.

In this discussion, comments are made regarding the model
methodology, spatial representation, adaptation problems of a general
model, data problems of a new application, and sensitivity analysis.
An example is used in each case to illustrate the point being made.

MODEL METHODOLOGY

In a recent Lake Erie study, we have had the opportunity to review numerous water quality models of aquatic ecosystems [Lorenzen and Chen, 1974]. We found that almost everybody was doing the same thing. The majority of the models reviewed are based on two fundamental principles.

1. That there is conservation of mass even though a constituent is changed by reactions from one form to another in the aquatic ecosystem, the Law of Conservation of Mass, and

2. That the rate of change of any constituent can be represented by a function of coefficients and one or more constituent concentrations that interact to cause the change, the Kinetic Principle.

All models try to represent the aquatic ecosystem by a series of interconnected segments. For each segment, there are the associated concentrations of water quality parameters, e.g., dissolved oxygen (DO), plant nutrients (CO_2, PO_4, NH_3, NO_4, and SiO_2), algae, zooplankton, and fish. The objective of the model is to follow or to simulate how these quality parameters change with time at each segment.

The procedure is first to identify important physical, chemical, and biological processes that may act to change the concentration of the water quality parameters. Table 1 shows the important processes generally considered.

The next step is to express the rate of concentration change associated with each process according to the kinetic principle. The kinetic expressions are summed and equated to the changing rate of mass in the segment, according to the law of conservation of mass.

Table 1.   Important Ecological Processes for Modeling

1. Physical Processes

    a.   Advection between segments

    b.   Diffusion between segments

    c.   Sedimentation from the segment

    d.   External input to the segment

    e.   Output to external from the segment

    f.   Reaeration

    g.   Solar insolation

2. Biochemical transformation, uptake, and release associated with the following:

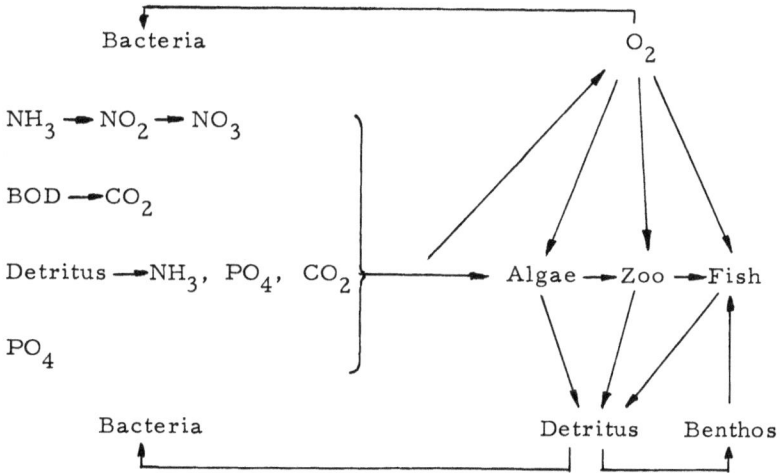

The resulting equations, commonly termed the mass balance equations, are solved numerically in both time and space to follow the ecological changes due to the influences of hydrology, waste input, meterology and other factors.

By taking only a "point" or a square meter of water column to represent a lake and by requiring temperature to be specified, CLEAN has ignored many important physical processes. CLEAN emphasizes biological transformations and uptake and release. Oxygen has not been simulated although oxygen is an important factor in determining fish survival or if bacteria are aerobic or anaerobic. Due to the model's inability to express nutrient stratification with water depth, CLEAN presumably uses average nutrient concentrations in the computation of algal growth rates. Knowing the non-linear nature of the growth expression, the procedure is far from satisfactory.

CLEAN, as the authors have claimed, does have a detailed representation for the complex biological system through the higher trophic levels. It is realistic for lakes where vertical stratification of temperature and plant nutrients are not severe. Oxygen throughout the lake depth must be high so that it is not a limiting factor for biological welfare.

SPATIAL REPRESENTATION

There are many ways to specify the spatial representation of aquatic ecosystems. A relationship exists between the detail of spatial representation and the time step of computation. Such a relationship exists not because of the numerical stability of the computation method. Rather, it is based on ecological considerations.

The simplest representation of a lake is to consider it as a completely mixed tank. For such a coarse representation, the model can not make an adequate simulation of seasonal algal growth dynamics. The model can only be used for a simple nutrient budget computation using a year as the time step.

Lorenzen [1973] and others [Imboden, 1974], have used such a procedure for the computation of phosphorus concentrations in a lake. The mass balance equation of the model considers the mass inflow, mass outflow, mass loss to sediment, and mass gain from sediment release. The annual loss to sediments is proportioned to the mean phosphorus concentration in the water and the gain from sediment is proportional to the phosphorus content of the sediment. The model has been shown to calculate the phosphorus concentration of Lake Washington for thirty-five (35) years, spanning the periods of accelerated cultural entrophication and sewage diversion. A preliminary computation has demonstrated its applicability to Lake Erie with a three-reactor representation. The question is of course how to relate phosphorus concentration to algal density. Such relationships do exist in Lake Washington [Edmondson, 1972], Lake Erie [Brydges, 1971], and undoubtedly numerous others.

A more detailed segmentation is necessary for the dynamic simulation of aquatic ecosystem. For rivers, it is a simple matter to cut the river into segments. For an estuary, a link-node system capable of two dimensional representation of an embayment is shown in Figure 1. Figure 2 shows the adaptation of such system to San Francisco Bay [Chen and Orlob, 1972].

For lakes, two layers, three layers [Lombardo, 1971, Thomann, 1973], and multiple layers have all been used for weekly computations. Chen and Orlob [1972] employ a stacked layer as thin as one meter to represent a stratified lake. This system requires a time step ranging from hourly to daily. For a long and

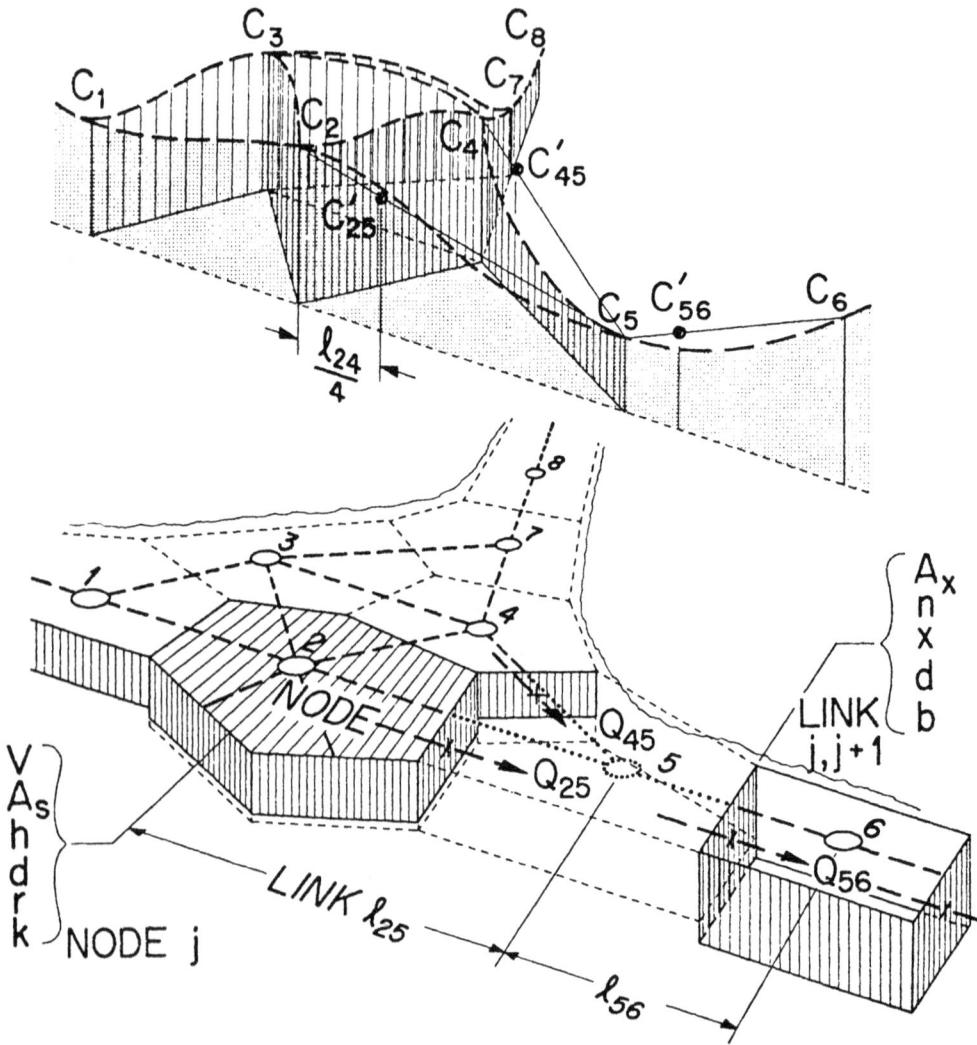

Figure 1   Typical Section of an Estuarial Node-Link Network

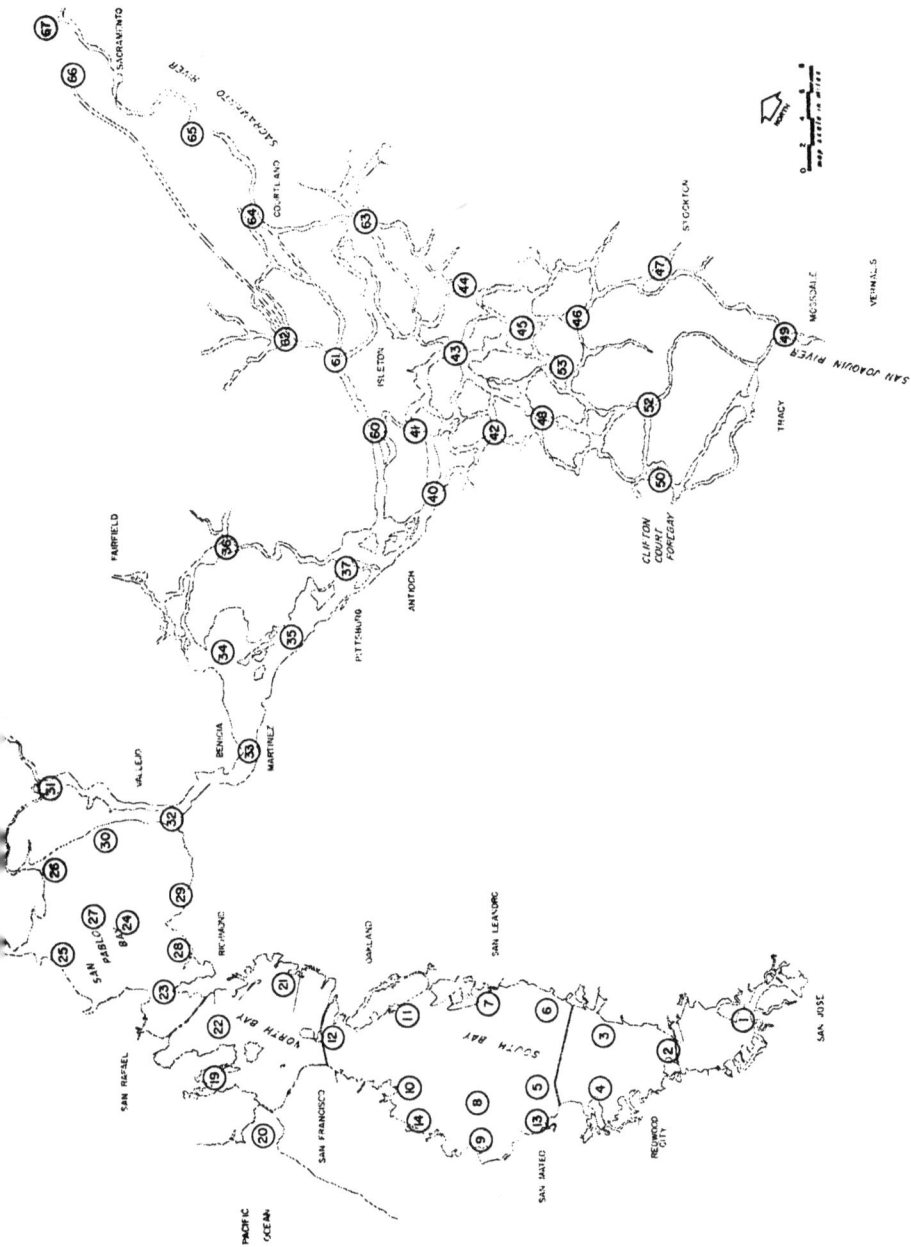

Figure 2   San Francisco Bay-Delta System, Nodal Point Representation

narrowed reservoirs, Bacca, et al [1974] adapted a system which divides the lake into reaches. Each reach is further divided into stack of horizontal layers.

For a very large lake, Canale [1973] proposed to divide Lake Michigan into five horizontal sections (see Figure 3). A narrow strip along the shore is for the littoral zone. The central portion of the lake is represented by two large zones. In the vertical direction, the lake is sectioned into three layers (see Figure 4).

Based on Canale's concept, it is proposed that Lake Erie be represented by a grid system shown in Figure 5 [Lorenzen and Chen, 1974]. Such a system will have a detailed horizontal spatial description for the near shore zone where pollutional effects are most severe. In the central core of the lake where variation in the vertical direction is more important, a large horizontal segment with numerous vertical layers may be appropriate.

The use of variable grid for the physical representation of a large lake is a significant conceptual development. First of all, it will be consistant with the spatial distribution of prototype water quality data. More importantly, it reduces greatly the number of segments necessary for a very large lake. The computer core size requirement can be within the reach of current computer hardware.

A few words should be mentioned about the computational requirements as a result of spatial segmentation. For Lake Washington, it takes 3.5 minutes of UNIVAC 1108 to perform an annual simulation with a daily time step and a 65 element (1 meter each) representation of the lake. In this simulation, hydraulic as well as twenty three (23) water quality parameters are calculated simultaneously. The cost is not too high. It should be considered as a necessary evil for the analysis of a complex ecological system.

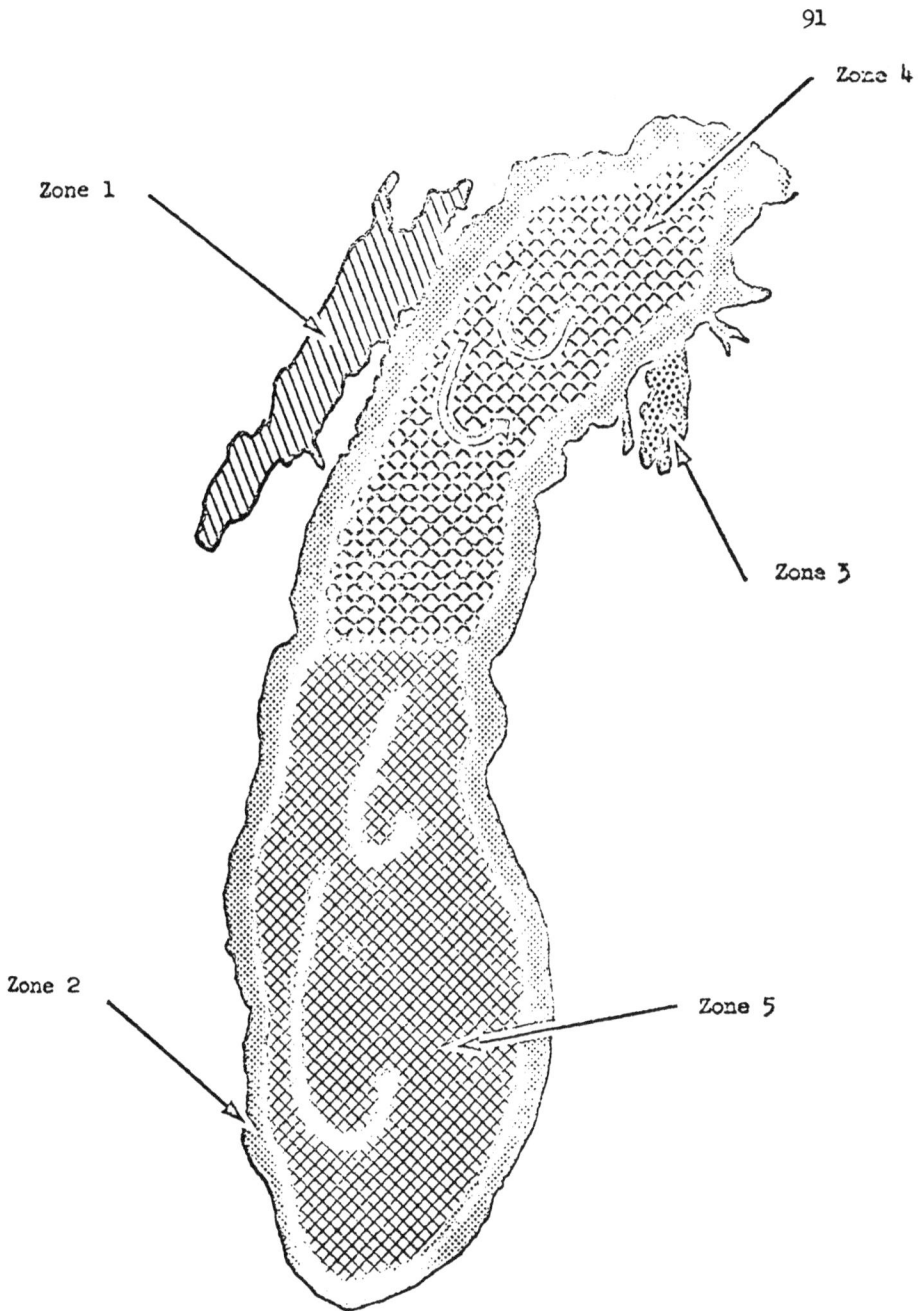

Figure 3   Hypothetical Horizontal Sectioning of Lake Michigan
             Into Uniform Zones (After Canale)

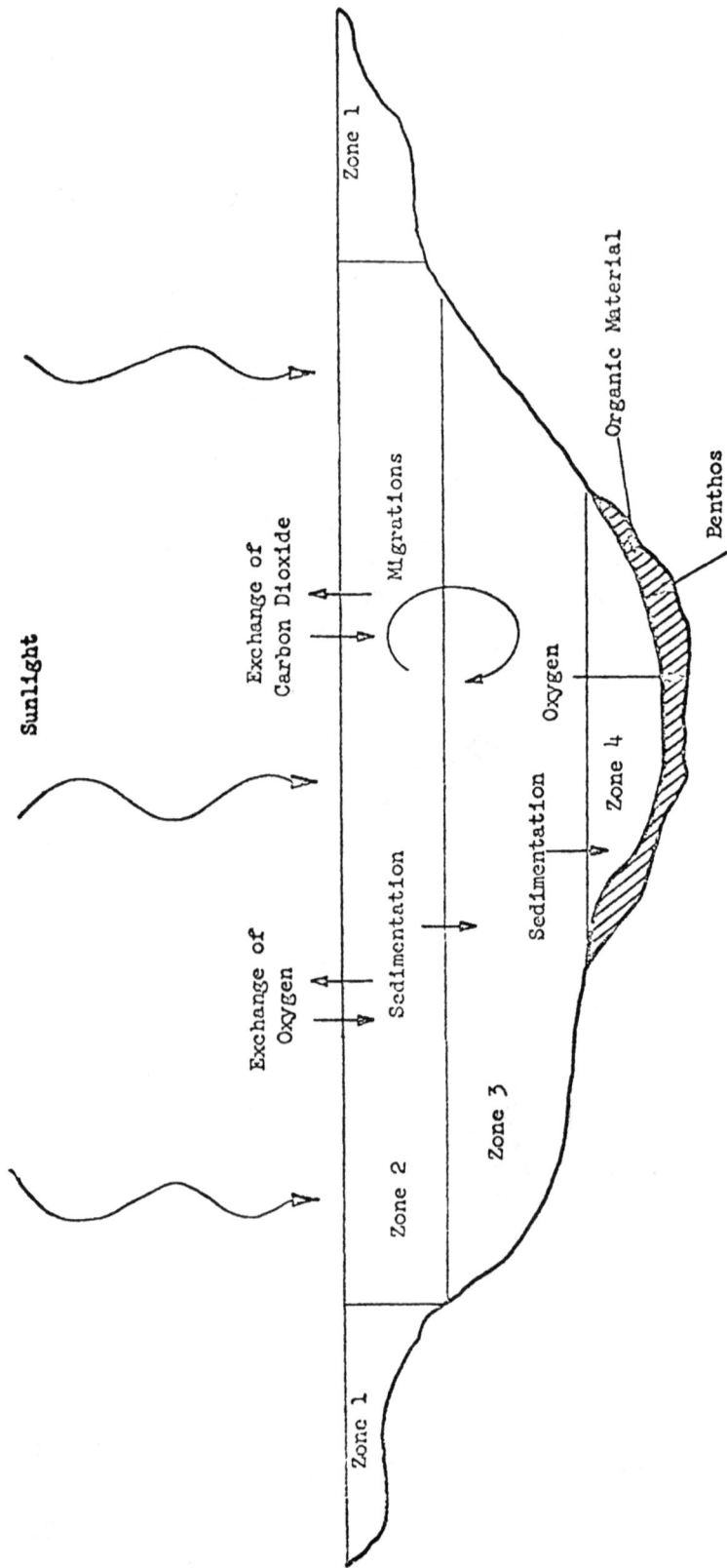

Figure 4   Vertical Sectioning of a Lake Into Uniform Cells   (After Canale)

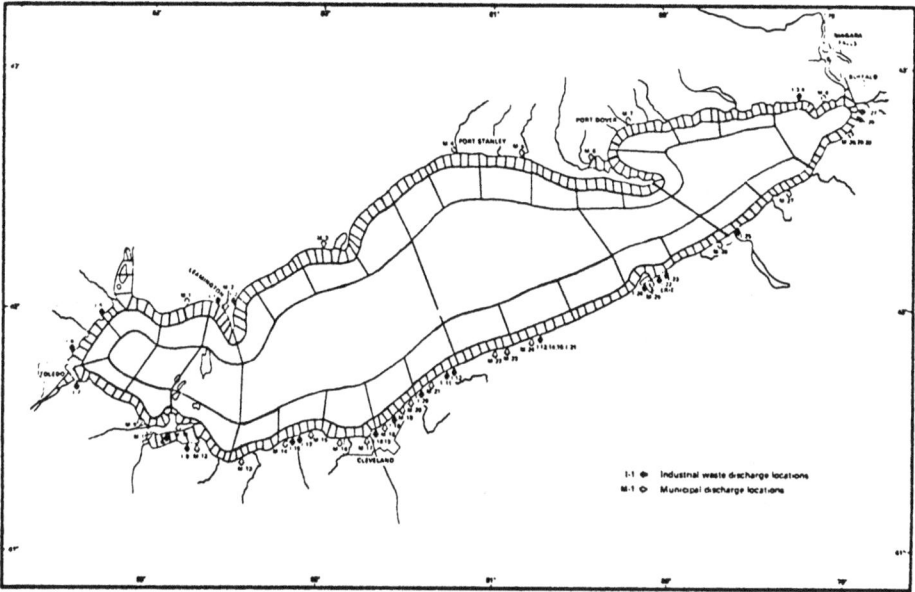

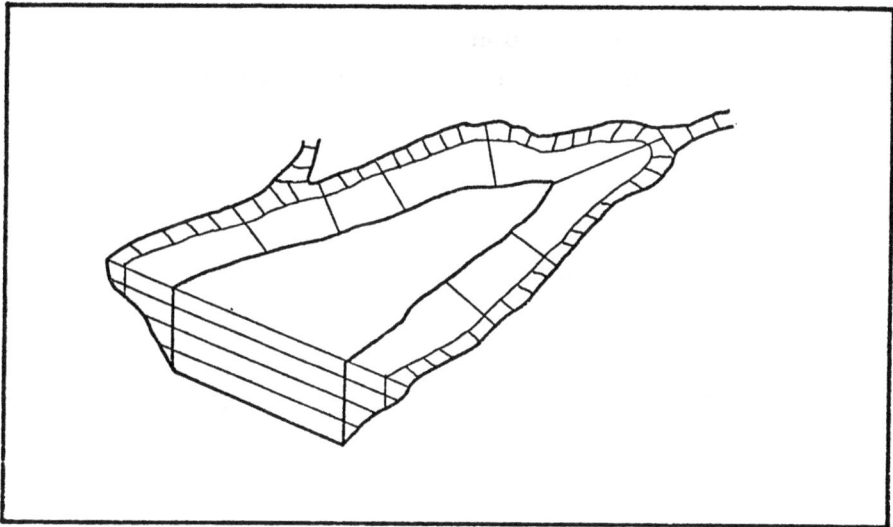

Figure 5   Segmentation for Lake Erie

ADAPTATION OF A GENERAL MODEL

It is always a dream for investigators, including myself, to develop a general model usable for many lakes. In our experience, however, we find that there is always something to be modified in a new application.

When the model developed for Lake Washington was applied to Lake Koocanusa-Libby Dam in Montana, we had to develop an ice formation routine in the heat budget computation and also to include suspended solids in the computation of fluid density in addition to the temperature effect [Chen and Orlob, 1973]. To apply such a model to Lakeport Lake which is a two branch reservoir, the program was modified to treat the system as two reservoirs with interflows [Chen, Lee, and Lorenzen, 1974]. A pumped storage application is another example of modification [Chen, 1972].

A straight application of the model to some warm water lakes in Texas created problems as well. Temperature adjustment functions for most rate coefficients were found abnormal for a lake reaching $35^{\circ}C$ on the surface. In a twenty year continuous simulation of a lake, nitrogen was found to be accumulated in the system, not consistent with the observed data. It was later found that the lake bottom went anaerobic during the summer. Denitrification must have occured during the summer. Implementation of a denitrification process in the model produced good results.

A model that has not been applied to more than three (3) prototype conditions can not be considered general. Through applications, a model can be improved and made more general.

It is true, as noted by the authors, that some modifications can easily be made on the computer program. However, the work does not stop there. To underestimate the man-hour requirement for a modification is a dangerous act of enthusiastic modelers (myself included).

DATA PROBLEMS

In many practical applications, we do not have the luxury of a large interdisciplinary research team for data collection. In some lakes where there are in-lake water quality data, there is no waste input or nutrient load information. Data are not only sparse, but also the wrong kind. After many years of work in this field, one almost develops a feeling that the need and desire for a model study of a lake is inversely related to the availability of the data.

One example of such a situation is the preimpoundment study. In such a study, weather data are obtained from a nearby station. Hydrology is secured from the U. S. Geological Survey or other Water Resources Agency. The determination of inflow water quality usually requires a new or supplemental sampling program. The inflow qualities are correlated with flow quantities for the estimate of qualities for the ungaged period. System rate coefficients are compiled from typical values reported in the literature. Results are compared with the water quality data of the nearby reservoirs.

SENSITIVITY ANALYSIS

The authors are very correct in stating that the model can help understand the aquatic ecosystem. One may add that the insight gained about the ecosystem behavior is the primer for the formulation of management alternatives or research problems related to the aquatic environment. The solution comes from the understanding of system behavior rather than from the charts showing the exact reproduction of observed data by the model.

Understanding of ecosystem behavior can be obtained by sensitivity analysis. Many types of sensitivity analysis can be performed. Thomann, et al [1972] have conducted a sensitivity analysis with a simple lake model. Lake Ontario was represented by three layers. The model included one species of phytoplankton and two species (one herbivore and another carnivore) of zooplankton. They were able to compute two blooms of algae, implying that species differentiation of phytoplankton population ought not be important or responsible for the bimodel algal population observed in the Great Lakes.

Chen, et al [1974], on the other hand, have applied the lake ecologic model to Lakeport Lake, California. The model has two species of algae with different nutrient, light, and temperature requirements. They also settle at different rates. A single species of zooplankton grazes two algal groups with preference. During the simulation, it was found that the lake would develop bimodal algal population during the dry year. Further analysis revealed that nutrient income during the dry year was low.

Thus, an oligotrophic lake will support many small blooms because of the low nutrient concentrations. An eutrophic lake for its abundant nutrient will continue to support the growth of the first species even after the second species picks up its growth rate in late summer. There may be a population shift from the first species to the second, however, the total phytoplankton biomass shows only one peak in a year.

In the same study for Lakeport Lake, we have varied the growth rate and half saturation constants for Algae 2 which is to symbolize blue-green. First of all, we found that these values could not be changed arbitrarily. Within the range allowed, the simulation results indicate a small likelihood for blue-green to pro-liforate in the lake. The main reason is that the lake is more deficient in phosphorus than nitrate. There is no edge for blue-green algae due to its nitrogen fixing capability.

Many sensitivity analysis were made with Lake Washington simulations. Several interesting insights were obtained. In order for the model to predict a pH of 9.2 in the summer, the half satu-ration constant for carbon must be reduced from 0.5 mg/l to 0.03 mg/l and the reaeration coefficient of carbon dioxide must be re-duced to ten percent (10%) of reaeration coefficient for oxygen. [Chen and Smith, 1974]. The low value of the half saturation con-stant for carbon has, incidently, been supported by recent experi-mental results of Jenkins, et al [1974].

In a previous study, Chen and Orlob [1972] reported that the lake model predicted a less drastic improvement of the Lake Washington ecosystem than the one actually observed after sewage diversion. It was speculated that the estimate of initial conditions for the diversion case did not take into considerations the waste load reduction. In the recent sensitivity analysis, the initial con-ditions in January were halved. The model responded with a drastic reduction in algal density and also a marked improvement in the hypolimnion DO.

One may argue that the results are as expected. However, it is intriguing to observe by model simulation that Lake Washington is completely mixed in the winter. As the water becomes stratified in the summer, most of the nutrient ladden inflows enter the lake hypolimnion. The contemporary nutrient income is therefore not available for algal growth on the lake surface. In essense, the epilimion water is practically sitting on the lake surface waiting for the conversion of all nutrients to algal biomass. The nutrients at the surface are replenished by mixing that occurs during the fall overturn.

There are some individuals who may say that sensitivity analysis is a way for modelers to get whatever answers they want. I would suggest that sensitivity analysis is a powerful method to gain insight to ecosystem behavior.

References

1. Baca, R. G., Lorenzen, M. W., Mudd, R. D., and Kimmel, L. V., "Application of a Water Quality Simulation Model to American Falls Reservoir", Final Report to the EPA, Seattle, Battelle Northwest Laboratory, Richland, Washington, 1974

2. Bloomfield, J. A., et al, "Aquatic Modeling in the Eastern Deciduous Forest Biome, U. S. IBP Program", Workshop Proceeding, "Modeling the Eutrophication Process", Utah State University, Logan, Utah, November 1973

3. Brydges, T. G., "Chlorophyll a - Total Phosphorus Relationships in Lake Erie", Proc. 14th Conference, Great Lakes Res., 1971

4. Canale, R., "A Methodology for Mathematical Modeling of Biological Production", University of Michigan, Sea Grant Program, 1971

5. Chen, C. W. and G. T. Orlob, "Ecologic Simulation for Aquatic Environments", Report to the Office of Water Resources Research, Water Resources Engineers, Inc., Walnut Creek, California, 1972

6. Chen, C. W. and Orlob, G. T., "Predicting Quality Effects of Pumped Storage", Journal of the Power Division, ASCE, Vol. 98, No. PO1, June 1972

7. Chen, C. W. and Smith, D. J., "Model Studies of Puget Sound and Lake Washington", Report to Water Resources Engineers, EPA, and Office of Water Resources Research and Technology, Tetra Tech, Inc., Lafayette, Calif., 1974

8.   Chen, C. W., "Limnological Study for Helms Pumped Storage Project", Report to the Pacific Gas and Electric Co., Water Resources Engineers, Inc., Walnut Creek, California, 1972

9.   Chen, C. W. and Orlob, G. T., "Ecologic Study of Lake Koocanusa Libby Dam", Report to the Corps of Engineers, Seattle District, Water Resources Engineers, Inc., Walnut Creek, California, 1973

10. Chen, C. W., Lee, S. S., and Lorenzen, M. W., "Water Quality Studies for the Proposed Lakeport Lake", Report to the Corps of Engineers, Sacramento District, Tetra Tech, Inc., Lafayette, California, 1974

11. Edmondson, W. T., "Nutrients and Phytoplankton in Lake Washington", in Nutrients and Eutrophication, American Society of Limnology and Oceanography, 1972

12. Imboden, D. M., "Phosphorus Model of Lake Eutrophication", Limnology and Oceanography, Vl19, No. 2, March 1974

13. Lombardo, P. S., et al, "Mathematical Model of Water Quality in an Impoundment", M. S. Thesis, University of Washington, Seattle, Washington, 1971

14. Lorenzen, M. W. and Chen, C. W., "Lake Erie Wastewater Management Study", Final Report to the Corps of Engineers, Buffalo District, Tetra Tech, Inc., Lafayette, California, October 1974

15. Lorenzen, M. W., "Predicting the Effects of Nutrient Diversion of Lake Recovery", Workshop Proceeding, "Modeling the Eutrophication Process", Utah State University, November 1973

16. Park, R. A., et al, "A Generalized Model for Simulating
     Lake Ecosystems". A reprint paper accepted for publication
     in SIMULATION, August 1974.

17. Thomann, R. A., et al, "Mathematical Modeling of Eutrophi-
     cation of Large Lakes", First Annual Report of the EPA
     IFYGL projects on Lake Ontario, December 1973

THE DELAWARE ESTUARY

Robert A. Kelly

I. Introduction

With the current interest in ecological modeling as
evidenced by several recent books (Smith, 1974; Patten, 1971,
1972; EPA, 1971; Pielou, 1969), and this Symposium I might
add, it appears that the time is now approaching where relatively sophisticated ecological models may be used in resource management decision making. Although the level of
understanding of ecological systems is still very low, many
general principles and empirical observations on various systems allow construction of relatively realistic simulators
of the natural world. The incorporation of these simulators
into a specific management framework then allows analysis of
various management alternatives to achieve desired objectives.

This paper concerns the development of an ecological model which attempts to integrate several different aspects of water quality into a single framework, in order to be able to quantify the relative contribution of discharges of various pollutants at selected points to violations of water quality standards wherever they occur. The beauty of the construct is that the tradeoffs between reducing the discharges of all materials and reducing only the discharges of materials which are contributing to standards violations can be explicitly evaluated. This will be explained more fully later.

II. The Delaware Valley Model

The Delaware Valley

The Quality of the Environment Program at Resources for the Future has, for the last few years, been involved in the development of a residuals (waste) management model for the lower Delaware Valley. The area involved (Figure 1) is one of the most developed industrial areas in the country, with 7 petroleum refineries, 5 steel mills, and several other significant industries; has hundreds of sewage treatment plants serving many small communities and several major urban areas of which Philadelphia-Camden is the largest; and is called home by over five million people. According to federal air quality standards, these millions are inhabiting an area of poor environmental quality, and it appears that even if an immense air pollution control program were instituted, the quality of air prescribed by the Environmental Protection Agency could not be attained. On the aquatic side, the Delaware Estuary serves as the receptacle for over 500 tons of organic matter per day, plus

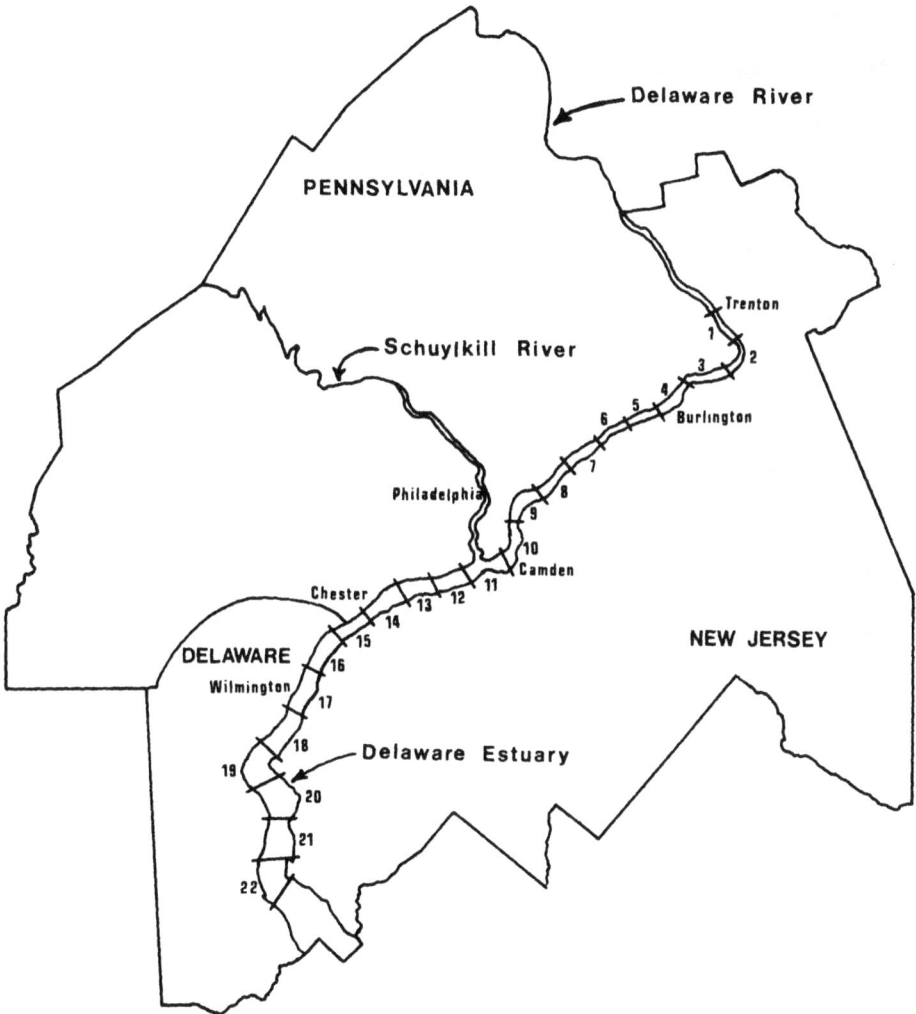

Figure 1. Map of area of concern for the Delaware Valley
Regional Management Model, showing reach locations for the estuary model.

thousands of pounds of nitrogen, phosphorus, suspended solids, and toxic materials. By standards set by the Delaware River Basin Commission (DRBC), water quality is poor. However, unlike the air, the water quality standards can apparently be met using existing technology and at relatively low cost.

The Residuals Management Model

The residuals management model is composed of three parts: (1) an industrial and municipal residuals generation, treatment, and discharge submodel--a linear programming model of several major industries and other waste generating activities; (2) an environmental model section--an air dispersion model and an aquatic ecosystem model; and (3) an environmental evaluation section--an algorithm which translates violations of ambient environmental standards to "marginal penalties," which reflect the marginal contributions of each discharge to ambient standards violations. The details of this model are outlined in two papers (Russell et al., 1974; Spofford et al., 1972) with considerable background and explanatory material appearing in Russell and Spofford (1972).

III. Ecological Model of the Delaware Estuary

Although the Delaware Valley Residuals Management Model (DVRMM) as a whole is a fascinating and complicated machine, the linkages between the ecological model and the rest of DVRMM are of more immediate interest. The linkages are discharges of materials (amount per day at a specified location) and marginal penalties ($ per unit discharge per day) which are placed on each discharge according to each discharge's

contribution to ambient standards violations. For each discharge type
and location in the residuals generation model, the environmental models
must have a corresponding input in type and location, and the environmental evaluation model must be able to generate marginal penalties for
that type and location. In the DVRMM, the types of discharges include
the following: biological oxygen demand (BOD), nitrogen (as total Kjeldahl nitrogen--organic N plus ammonia), phosphorus (elemental), suspended solids, heat (Btu's), toxic materials, and oxygen (through in-
stream aeration). The spatially distributed discharges of these materials are considered exogenous variables to the ecosystem model.

The desired water quality variables which are endogenous variables
in the ecosystem model are the concentration of algae, the biomass of
fish, and oxygen concentration. These variables are not usually used as
water quality standards parameters (with the exception of oxygen) but
are included to demonstrate the possibility of explicitly incorporating
within the model things which people might consider meaningful.

It is somewhat misleading, however, to enter into a discussion of
the structure of the ecological model in this way. The endogenous variables were not determined at the outset of the project. As time passed,
discussion among the interdisciplinary team occurred which caused modifications to be made in both the industrial models and the ecological
model. The types of residuals discharged were set fairly early due to
limitations in data availability for the industrial processes involved,
and the ambient environmental standards were picked after the ecosystem
model began to take shape. For some variables (e.g., coliform bacteria),
a simple modification of the ecosystem model would allow their

incorporation as water quality standards, but since these variables would be treated as if a separate model and not strongly interacting with the rest of the ecosystem, they were not included.

Objectives

The design of the ecological model at the beginning of its development was guided by the following considerations:

1. The model should be able to handle orders of magnitude differences in inputs of waste materials. Although all models can deal with this in principle, it seemed to me that only a nonlinear representation could realistically include the types of responses that might be exhibited under such different conditions.

2. The model should include, as far as possible, the known major pathways of natural flows within the ecosystem. Rates of materials transfers could then be compared with literature values in order to test their validity.

3. Only published or easily available data could be used. No new field data could be obtained.

4. The model was to be used in conjunction with a static economic model and thus should be steady state rather than a time simulation.

Variables

The endogenous variables for the model were chosen on the basis of available data, presumed functional importance of types of organisms, residuals inputs, and the necessity for management handles. They are algae, zooplankton, bacteria, fish, oxygen, nitrogen, phosphorus, toxics concentrations, suspended solids, and temperature.

The earliest attempts at ecological models usually included only algae and zooplankton (Patten, 1968). The reasons for including other variables are given below.

Bacteria are the decomposers of the system. In the classical consideration of BOD, the nature of the organic matter and bacteria determine the decay rate, k_1 (Fair and Geyer, 1956). Since the nature of the kinetics of bacterial uptake of substrates and respiration is well understood in general, the conversion of BOD and the respiration of bacteria is uncoupled so that there is no stoichiometric relationship between the two. However, only one study that I am aware of has been done on the natural (as opposed to coliform) bacteria in the Delaware (Glaser, personal communication), and any quantity of data is sorely lacking.

Fish are included in this model solely as an ambient water quality standards output. Their role in determining behavior of the model is limited, and their reason d'etre in the ecosystem proper is unclear. Ichthyological Associates (1970, 1969, 1968) and Abbe (1967) have done research on the numerical distribution of fish, but mass of fish at any point in time and space is unknown. However, as with the bacteria, their behavior in the environment as a function of toxic materials concentrations, oxygen concentration, food supply, and temperature is known in general, and they are included in the Delaware model to indicate trends with few claims for accuracy.

Algae were included as the base of the food chain, as an indicator of a potential water quality problem, and as an important factor in the oxygen balance of the estuary. Their importance in the food chain appears to be at the lower end of the estuary (and to an unknown extent at the upper end) where allochthonous sources of energy for the rest of the biota are relatively small. Algae are not now a serious water quality problem, but could be if an effort were made to reduce the turbidity (suspended solids) load which the estuary now carries. Even then, it might be only a temporary problem in the spring (Thomann, personal communication). Finally, the oxygen production of algae is apparently very important in the upper and lower parts of the estuary. At these places, the 24-hour variation in oxygen concentration can be as high as five milligrams per liter (USGS, 1968).

Model Structure

The differential equations which describe the time behavior of these biological components are closely coupled with each other and with the equations for nutrients, oxygen, and BOD. The extent of the coupling is partially indicated in Figure 2. The rate of change of any component is dependent on the interrelationships between feeding (nutrient uptake for algae), respiration, excretion, and death of each of the living components. Feeding, respiration, etc., are represented by separate terms within the differential equations describing component behavior. The rationale for each term is given below.

The feeding rate is a function of the amount of consuming biomass, the amount of biomass available for consumption, the temperature, the oxygen concentration, and the concentration of toxics materials. The effects of each of these factors are my guesses as to how the various processes are affected, and their nature is indicated graphically in Figure 3. Each effect could profitably be the subject of several research projects.

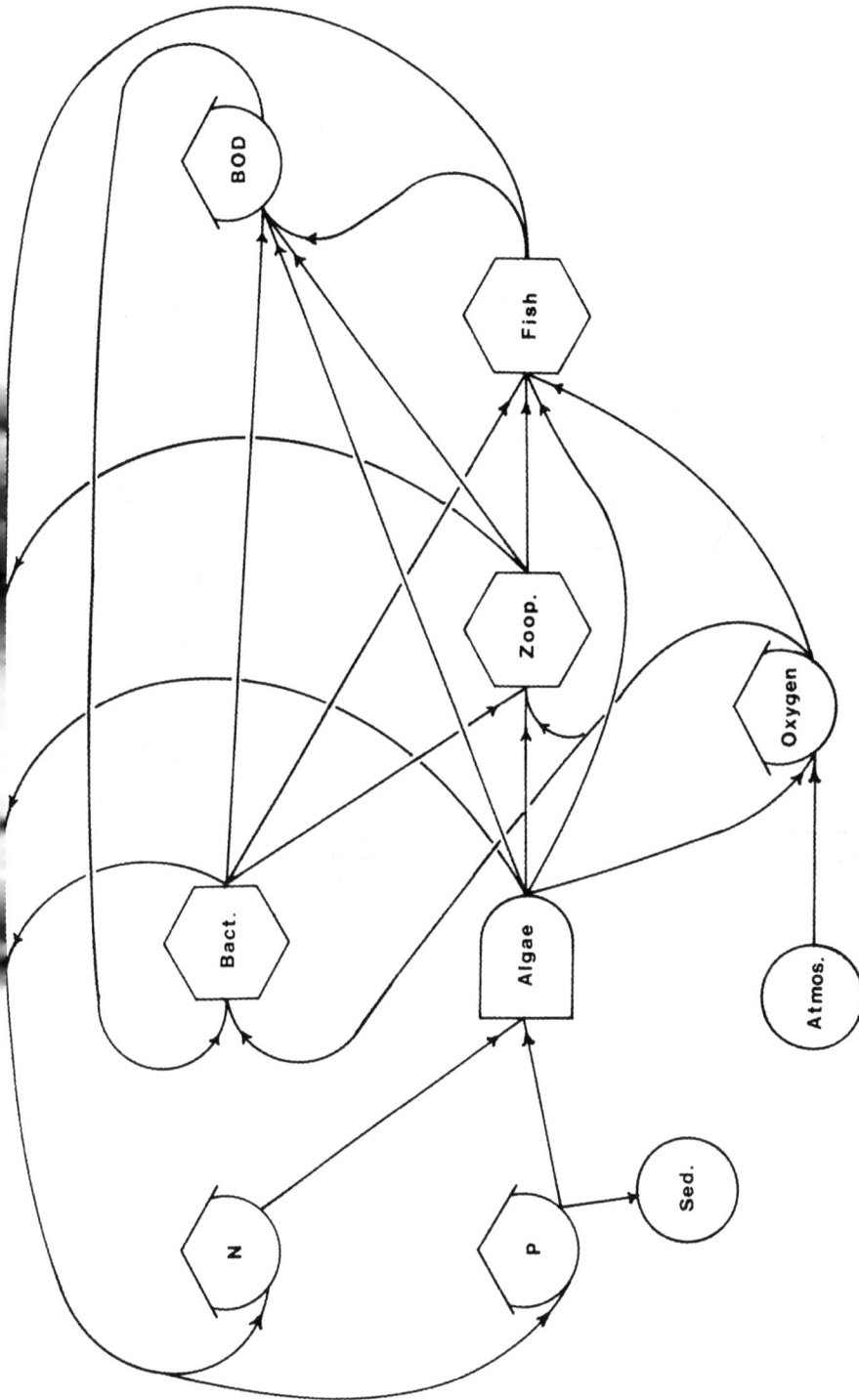

Figure 2. Diagram of material flows within the ecosystem for one reach.

112

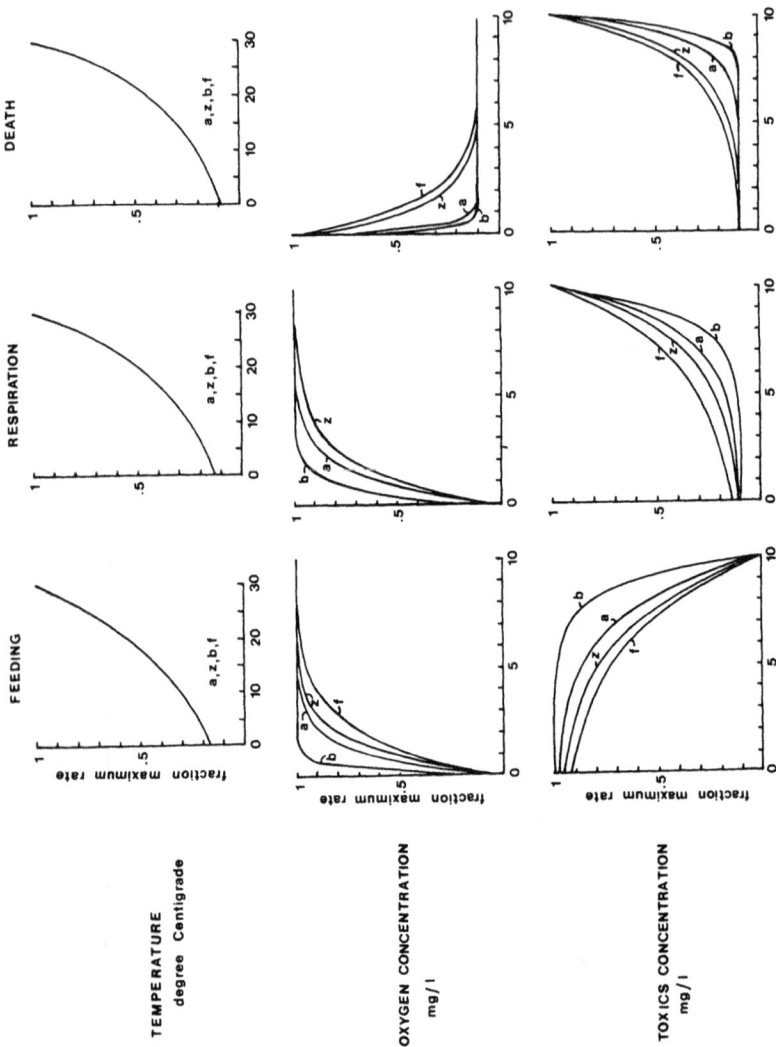

Figure 3. Relative effect on various rates of environmental parameters. (a = algae, b = bacteria, z = zooplankton, f = fish)

Respiration is a function of the biomass consuming oxygen, the temperature, the oxygen concentration, and the concentration of toxic materials. The shapes of the curves in Figure 3 (with the exception of toxic materials) are fairly well documented in the literature for a large variety of systems (see, for example, McDonnell and Hall, 1969; Owens and Maris, 1964; Silver et al., 1963; Herrman et al., 1962). However, the nature of the interrelationships between factors is not known at all (in the sense of antagonistic or synergistic effects). Here, as with other processes, effects are assumed to be multiplicative.

Excretion is the most unrealistically handled process in the ecological model. It is assumed to increase as the square of the concentration of biomass. A more realistic assumption (without knowing the effect on component behavior) would be to increase excretion proportional to the availability of food. However, this is apparently not the case for algal excretion (Anderson and Zeutschel, 1970) where higher excretion rates of organic compounds per unit biomass occur in oligotrophic environments than in eutrophic environments.

It is assumed that natural death (old age) is very rare in an ecosystem and that death is a function of oxygen concentration, temperature, and toxics concentrations.

Four components of the ecological model have special processes which are physical-chemical in nature. Two of these were included after observing preliminary behavior of the model, two were included prior to the first runs.

Phosphorus and suspended solids precipitate or settle to the sediments from the water column under appropriate conditions. Both of these materials are assumed to be lost from suspension at a rate proportional to their concentration and inversely proportional to the reaeration rate.

The rate of reaeration is assumed to be proportional to the difference between actual oxygen concentration and the saturation concentration of oxygen for specified conditions as in standard engineering practice.

Only total Kjeldahl nitrogen is considered in the model, and early runs indicated that organic nitrogen and ammonia must be transformed to other forms of nitrogen in the estuary (see, e.g., Kelly, 1974). The conversion of ammonia to nitrate (nitrification) as a function of oxygen concentration, temperature, and toxics concentration caused the model to behave more realistically with respect to nitrogen concentrations measured in the estuary. However, the conversion was not coupled with the oxygen system for two reasons: first, the oxygen concentration predicted by the model seemed adequate without the nitrification term, and inclusion of this term would only lead to a worsening of the fit between predicted and observed oxygen concentrations; and second, the amount of oxygen consumed during nitrification is not necessarily stoichiometrically related to the amount of nitrogen converted as is always assumed. Although the general consensus appears to be that 4.57 grams of oxygen are consumed per gram of ammonia converted, and that heterotrophic nitrification is very rare (Tuffey, personal communication), I disagree with the consensus.

Toxic materials represent a conglomeration of all types of pesticides, phenols, cyanides, and other materials lumped into one variable.

Ideally, because of the various effects each specific compound can have, each should be considered separately. Lack of data on concentrations in the estuary as well as lack of knowledge on physiological effects of poisons forces their inclusion here as a variable solely for interaction with the industrial models. As such, toxics are assumed to decay at a rate proportional to their concentration. It is interesting that behavior of the model is rather strongly dependent on the distribution of toxics throughout the estuary.

The differential equations developed for and coefficients used in the model are presented in Figure 4.

Many current ecological models have concentrated on the biological interaction to such an extent that physical transport mechanisms have been ignored. This is understandable when one considers the relative accuracy and state of development of hydrodynamic models versus that of ecological models (see EPA, 1972). However, it is obvious that the spatial distribution of various materials will determine the behavior of the ecosystem, and few systems are well mixed enough to permit the assumption that they are homogeneous. It is standard water quality modeling procedure to incorporate advective (and sometimes dispersive) transport mechanisms explicitly within models. In many cases, the advective transport of materials and a simple decay term seem to be adequate explanatory variables for most non-conservative wastes. This suggests to me that advective transport must be important.

The simplest way in which to include advective transport is to divide the estuary into several sections and to describe the flow of water (and thus dissolved and suspended materials) between the sections.

Figure 4. Differential equations and values of coefficients used for one section of ecological model

Symbols used:

f -- feeding
p -- predation
r -- respiration
d -- death
i -- input from industrial, municipal, or tributary sources

e -- excretion
t -- transport
α -- reaeration
β -- nitrification

γ -- precipitation
δ -- sedimentation
ϵ -- toxics decay
ζ -- heat exchange

Algae $\dot{A} = f_A - r_A - d_A - e_A - p_A + t_A$

Zooplankton $\dot{Z} = f_Z - r_Z - d_Z - e_Z - p_Z + t_Z$

Bacteria $\dot{B} = f_B - r_B - d_B - e_B - p_B + t_B$

Fish $\dot{F} = f_F - r_F - d_F - e_F$

Oxygen $\dot{O} = 2.319\left\{f_A - r_A - r_Z - r_B - r_F\right\} + \alpha + t_O + i_O$

BOD $\dot{L} = d_A + e_A + d_Z + e_Z + d_B + e_B + d_F + e_F - f_B + t_L + i_L$

Nitrogen $\dot{N} = 0.1087\left\{r_A + r_Z + r_B + r_F - f_A\right\} - \beta + t_N + i_N$

Phosphorus $\dot{P} = 0.02174\left\{r_A + r_Z + r_B + r_F - f_A\right\} - \gamma + t_P + i_P$

Suspended Solids $\dot{\sigma} = -\delta + t_\sigma + i_\sigma$

Toxics $\dot{\varphi} = -\epsilon + t_\varphi + i_\varphi$

Temperature $\dot{\theta} = -\zeta + t_\theta + i_\theta$

Figure 4 (continued)

$$f_A = \{6.0\} \{A\} \{1 - e^{-3.00}\} \{0.17146e^{0.0587790}\} \{1 - e^{-0.4(10-\phi)}\} \left\{\frac{1}{(0.151+N)(0.015+P)}\right\} \left\{\frac{2.66}{Z(.05+.17A+.10B+.05\sigma)}\right\}$$

$$f_Z = \{1.5\} \{Z\} \{1 - e^{-0.80}\} \{0.17146e^{0.0587790}\} \{1 - e^{-0.3(10-\phi)}\} \left\{\frac{1}{2}\left(\frac{1}{4.0+A} + \frac{1}{1.25+B}\right)\right\}$$

$$f_B = \{3.3\} \{B\} \{1 - e^{-1.00}\} \{0.17146e^{0.0587790}\} \{1 - e^{-0.9(10-\phi)}\} \left\{\frac{1}{7.0+L}\right\}$$

$$f_F = \{1.2\} \{F\} \{1 - e^{-0.60}\} \{0.17146e^{0.0587790}\} \{1 - e^{-0.25(10-\phi)}\} \left\{\frac{1}{3}\left(\frac{1}{3.0+A} + \frac{1}{2.0+Z} + \frac{1}{4.5+B}\right)\right\}$$

$$p_A = \frac{f_Z}{1 + \dfrac{4.0+A}{1.25+B}} + \frac{f_F}{1 + \dfrac{3.0+A}{2.0+Z} + \dfrac{3.0+A}{4.5+B}}$$

$$p_Z = \frac{f_F}{\dfrac{2.0+Z}{3.0+A} + 1 + \dfrac{2.0+Z}{4.5+B}}$$

$$p_B = \frac{f_Z}{\dfrac{1.25+B}{4.0+A} + 1} + \frac{f_F}{\dfrac{4.5+B}{3.0+A} + \dfrac{4.5+B}{2.0+Z} + 1}$$

$$p_F = 0$$

Figure 4 (continued)

$$r_A = \{8.0\}\ \{A\}\ \{1 - e^{-0.80}\}\ \{0.125e^{0.069315\phi}\}\ \{0.1 + e^{0.5(\phi - 10.2107)}\}$$

$$r_Z = \{2.3\}\ \{Z\}\ \{1 - e^{-0.60}\}\ \{0.125e^{0.069315\phi}\}\ \{0.1 + e^{0.4(\phi - 10.2634)}\}$$

$$r_B = \{4.5\}\ \{B\}\ \{1 - e^{-1.50}\}\ \{0.125e^{0.069315\phi}\}\ \{0.1 + e^{0.8(\phi - 10.1317)}\}$$

$$r_F = \{1.0\}\ \{F\}\ \{1 - e^{-0.50}\}\ \{0.125e^{0.069315\phi}\}\ \{0.1 + e^{0.3(\phi - 10.3512)}\}$$

$$d_A = \{1.0\}\ \{A\}\ \{0.1 + e^{4.0(\underline{Q} + 0.0263)}\}\ \{0.09391e^{0.078846\phi}\}\ \{0.1 + e^{1.0(\phi - 10.1054)}\}$$

$$d_Z = \{2.0\}\ \{Z\}\ \{0.1 + e^{0.9(\underline{Q} + 0.1171)}\}\ \{0.09391e^{0.078846\phi}\}\ \{0.1 + e^{0.6(\phi - 10.1749)}\}$$

$$d_B = \{3.0\}\ \{B\}\ \{0.1 + e^{5.0(\underline{Q} + 0.0211)}\}\ \{0.09391e^{0.078846\phi}\}\ \{0.1 + e^{3.0(\phi - 10.0351)}\}$$

$$d_F = \{0.5\}\ \{F\}\ \{0.1 + e^{0.7(\underline{Q} + 0.1505)}\}\ \{0.09391e^{0.078846\phi}\}\ \{0.1 + e^{0.5(\phi - 10.2634)}\}$$

$$e_A = \{0.04\}\ \{A\}\ \{A\}$$

$$e_Z = \{0.05\}\ \{Z\}\ \{Z\}$$

$$e_B = \{0.09\}\ \{B\}\ \{B\}$$

$$e_F = \{0.10\}\ \{F\}\ \{F\}$$

<u>Transport:</u>

1. For section i, dispersion (Thomann, 1972) included, for algae:

$$t_{A,i} = c_{i-1,i} A_{i-1} + c_{i+1,i} A_{i+1} - c_{i,i} A_i$$

2. For section i, dispersion omitted, for algae:

$$t_{A,i} = c_{i-1,i} A_{i-1} - c_{i,i} A_i$$

where $c_{i,j}$'s are functions of water flow, section volumes, and dispersion coefficients.

<u>Reaeration:</u>

For section i

$$\alpha = k_i (\underline{O}^* - \underline{O})$$

where k_i is the reaeration coefficient for section i, and $\underline{O}^*$ is the saturation concentration of oxygen at temperature θ.

<u>Nitrification:</u>

$$\beta = 0.1 \, N \, (1 - e^{-0.60} \underline{O}) \, (1 - e^{-0.9[10 - \phi]}) \, (0.17146 e^{0.058779\theta})$$

<u>Precipitation:</u>

For section i,

$$\gamma = 0.02 \, P/k_i$$

where k_i is the reaeration coefficient for section i.

<u>Sedimentation:</u>

For section i,

$$\delta = 0.02 \, \sigma/k_i$$

where k_i is the reaeration coefficient for section i.

<u>Toxics decay:</u>

$$\epsilon = 0.4 \, \theta$$

<u>Heat exchange:</u>

For section i,

$$\zeta = 0.683 \, ((\theta - \theta^*)/z_i$$

where z_i is the depth of section i in meters, and θ^* is the equilibrium temperature of section i (23.3C).

The number of sections to use involves several considerations including the homogeneity of the water mass, the variation in physical, chemical, and biological parameters, and the distribution of inputs along the length of the estuary. For the ecological model, the sections could be set up so that a three-, two-, or one-dimensional output could be obtained, but data limitations and computer time for solving more than one-dimensional systems made my decision to use a one-dimensional system more palatable. Indeed, previous work on the Delaware had used a one-dimensional approach, and the sections used for that model were employed with minor modifications (DRBC, 1970).

Division of the estuary into sections, however, is only part of the problem. How does one represent the flow of material between reaches? For a time-dependent simulation model, one needs water flow vectors between sections for each time step in the simulation procedure, requiring a relatively sophisticated (and expensive) hydrodynamic model. For a steady-state model, the mean of daily flows for a one-month-or-more period may be used, but this allows a choice of three methods. One can calculate the total volume of flow per day between each section (in both the upstream and downstream direction in an estuary) and use this in the model. A second approach is to use only the net flow between each section (which is always in the downstream direction) and thus ignore all upstream movements entirely. A third approach is to add a dispersion term to the net downstream flow and make it sufficiently large to account for the upstream movements of water as has been done by Thomann (1972). The effects of the latter two methods on output of the model are compared in Figure 5.

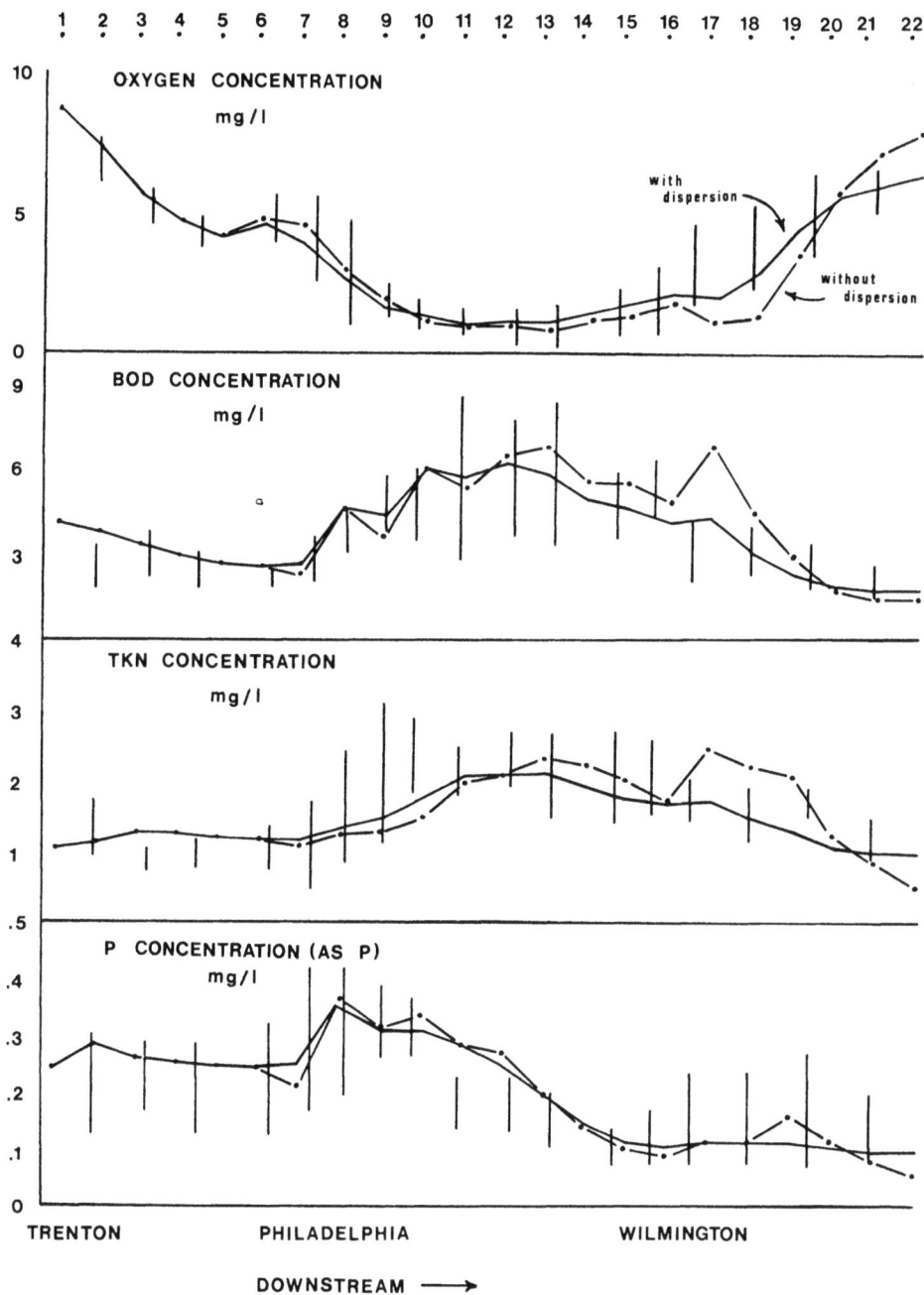

Figure 5. Comparison of model output with dispersion (solid line) and without (hatched line) to observed ranges of environmental parameters (vertical lines).

Figure 5 (continued).

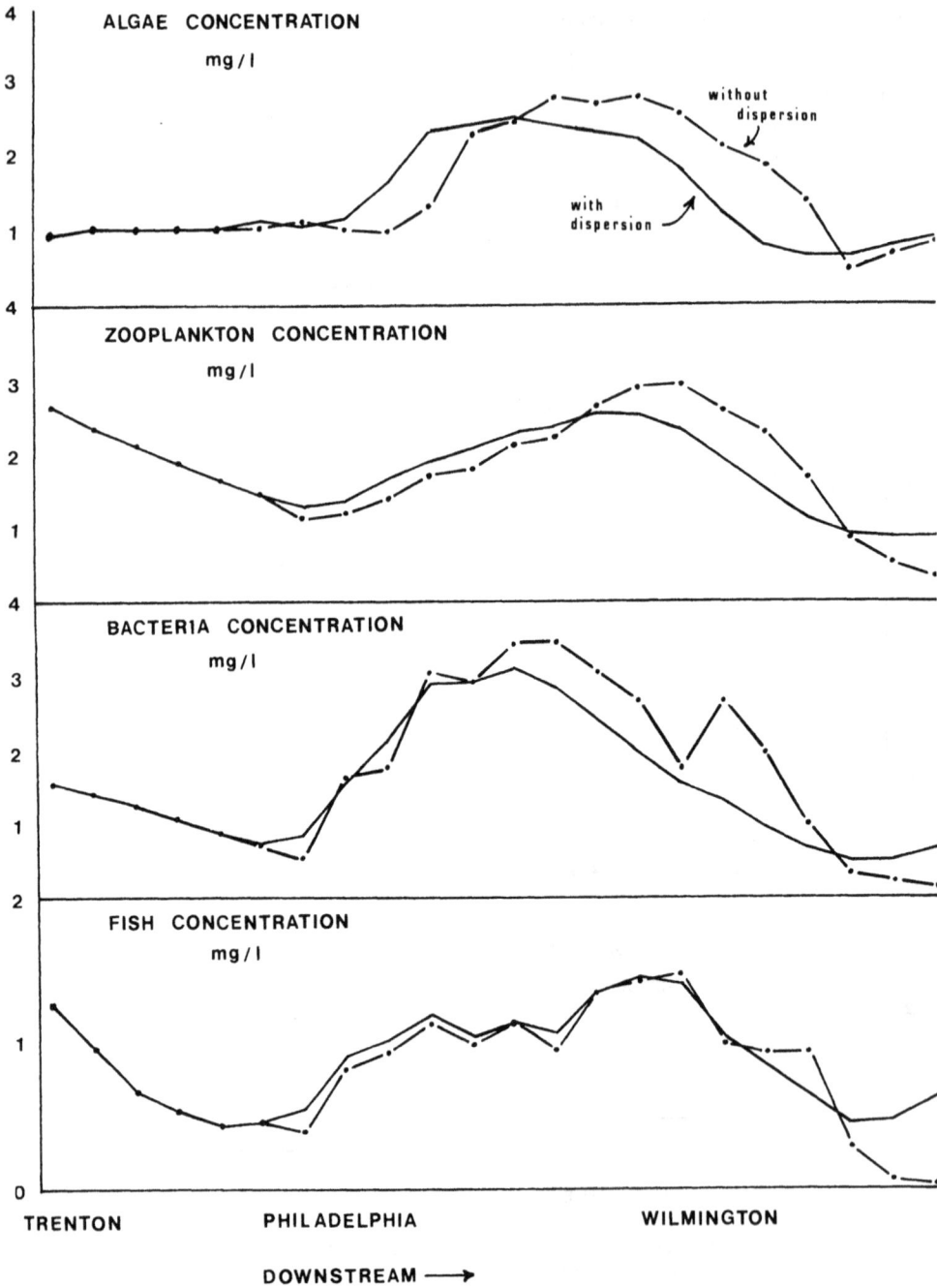

ALGAE CONCENTRATION
mg / l

without dispersion

with dispersion

ZOOPLANKTON CONCENTRATION
mg / l

BACTERIA CONCENTRATION
mg / l

FISH CONCENTRATION
mg / l

TRENTON   PHILADELPHIA   WILMINGTON

DOWNSTREAM ⟶

Omissions

Several important processes are not included in the model. Along the edges of the estuary, particularly the lower end, large marshy areas necessarily affect the behavior of the estuary with respect to nutrient sources and sinks and organic matter balance. These areas, although known in extent, are not understood in effect. They could be net importers or exporters of nutrients; contribute large amounts of BOD or take it away. They are presumably areas of high productivity and serve as sources of food for the rest of the biota in the estuary, but in what form and to what extent? Similarly, what happens in the sediments on the bottom of the estuary? For a large portion of its length, the bottom waters must be anaerobic, at least during the summer months. How does this affect the phosphorus balance? Are the organisms living there abundant or scarce? How much transfer of materials occurs between the sediments and water column? Are organisms instrumental in this transfer?

All of these questions are raised in an attempt to show that even though some aspects of the system are reproduced in the model, many other aspects are not. Even though the model behavior shows promise in its predictive capacility, is it doing so for the wrong reasons? For the purposes for which the model was built, all that is required is that it mimic nature in some way and that the results can be meaningfully interpreted. It poses interesting questions on the relative importance of certain processes and indicates at least the possibility of trying more complicated models in a predictive mode.

Results

As shown in Figure 5, most output variables follow the same trends as the data collected on the estuary for September 1970. A general comparison also indicates, understandably, that the model which includes dispersion provides a somewhat better fit to the historical data, especially at the lower end of the estuary where tidal influences are most pronounced. Unfortunately, adding dispersion as a descriptor of tidal flow adds complications to programming and solution time--it is much more efficient to solve each section for steady state individually from upstream and progressing downstream than it is to solve all sections simultaneously. The tradeoffs between computer time and accuracy (or realism) are not easily quantified.

It is apparent that there are a few problem areas with the model in comparing its output with September 1970 data. When using water flow and waste input data for May 1970 (where the major differences are in water flow and temperature), the problems are magnified greatly. Since there is almost no resemblance between model output and measured conditions for May 1970, no graphical comparison is necessary. Be that as it may, it is not understood why the conformance between model output and measured data is so poor for that month.

The data against which model output is compared is probably fairly good. For September 1970, the data were taken at five separate times, and although there is considerable variation in the data, general trends are apparent. The problem lies, then, either with the structure of the ecological model, the coefficients used in the equations, or the input data which is used to drive the model. Several of the problems with model structure have been identified above, and we need not linger on them. Values of the coefficients and input data remain to be treated.

Values of coefficients were chosen using a combination of hard data from the literature, intuition, and sheer guess. The Michaelis-Menton expression coefficients were chosen intuitively on the relative preference of food sources by a particular predator, or by examining literature values for algal uptake of nutrients. The relative effects of oxygen, toxics, and temperature on various rate processes of the different components of the ecosystem were determined by presumed relative sensitivities of different groups of organisms. All other rate coefficients were determined by experimentation with the behavior of the model, with rough limits on the relative magnitudes of values being set by published values of rates. Since there is no guarantee of a "true" set of values of the coefficients by this procedure, their choice could be offsetting improper model structure. The effect of the pattern of toxics, with high concentrations at the upper end of the estuary and low concentrations at the lower end, suggests that this may be so.

This is not to say that the input data is of good quality, however. EPA (1973) and the Delaware River Basin Commission (1970) have both published data on discharge from municipal sources to the estuary. For several sewage treatment plants, these discharges vary by as much as 50 percent for supposedly the same time period. Industrial discharges, as published in various sources, differ in a few cases by an order of magnitude. These data do not include the apparently small but significant and hard to measure inputs from tributaries and non-point sources (agricultural runoff, swamp drainage, and the like). Discharges from

these sources, when estimated by different methods, could easily vary
by a factor of five to ten.

So where does the problem lie? Only data collection on the estuary and on separate biological systems can tell. Only careful specification of problem areas and accurate measurement and synthesis will
improve our capabilities to model the Delaware ecosystem predictably.

IV. Ecosystem Response Matrix

A specific output was desired from the ecological model at the outset: what is the effect on a specified output variable in a specified
location of a change in amount of discharge of a specific type at a specific point? The answer can be obtained by one of two methods: numerically, by solving the model for different waste discharge values and calculating the change by difference; or calculating the response of the
ecosystem explicitly by analytically differentiating the equations. A
numerical approach was considered inappropriate because of lack of ac-
curacy--the differences in solution of the model with unit changes in
discharges were small enough to suspect that machine round-off and trun-
cation errors could be significant. This was traded against the problem
of developing the Jacobian matrix for the model and programming it--a
tedious task.

A large block of time was devoted to working out the following
matrix algebra which can be used to compute the ecosystem response matrix
(see Kelly and Spofford, 1974). Recently, in working with adding dispersion to the model, a mathematically simpler but computationally more
time-intensive approach became apparent. For easier referencing, I will
call one the section approach and the other the Jacobian approach.

Section Approach

The overall model can be expressed in one of several general forms. When tidal dispersion is ignored, a useful specification is

$$F_k = G(X_k) + Y_k \qquad\qquad k = 1,\ldots,22 \qquad\qquad (1)$$

where F_k is a vector of time derivatives for the endogenous variables for the kth reach; $G(X_k)$ are the rate expressions relating the endogenous variables to each other; and Y_k is a vector of inputs to each reach from upstream and downstream and waste discharges. From this general form, we ultimately need to develop a matrix of the form,

$$\left[\frac{\delta X_i}{\delta Z_j}\right] \qquad\qquad \begin{matrix} i = 1,\ldots,22 \\ j = 1,\ldots,i \end{matrix} \qquad\qquad (2)$$

describing the changes in endogenous variables in section i (X_i) with respect to changes in discharges in section j (Z_j).

For any section i.

$$\left[\frac{\delta X_i}{\delta Y_i}\right]$$

may be obtained explicitly by the relation

$$\left[\frac{\delta X_i}{\delta Y_i}\right] = \left[-\frac{\delta F_i}{\delta X_i}\right]^{-1} \cdot \left[\frac{\delta F_i}{\delta Y_i}\right] \qquad\qquad (3)$$

(see Sokolnikoff and Redheffer, 1958) where $\left[\dfrac{\delta F_i}{\delta X_i}\right]$ is the Jacobian matrix for section i, and $\left[\dfrac{\delta F_i}{\delta Y_i}\right]$ is the identity matrix (see equation (1)).

Since Y_i, in the model, is the sum of upstream inputs plus industrial discharges,

$$y_{i,j} = a_{i-1}x_{i-1,j} + c_i z_{i,j} \tag{4}$$

where $y_{i,j}$ is the input to section i of substance j; a_{i-1} is a transfer coefficient which is a function of water flow and section volumes; $x_{i-1,j}$ is the concentration of substance j in section i-1; c_i is a conversion factor for pounds day^{-1} to mgℓ^{-1}day^{-1} for section i; and $z_{i,j}$ is the discharge of material j to section i. Thus,

$$\left[\frac{\partial Y_i}{\partial Z_i}\right] = \begin{bmatrix} c_i & & & & \\ & c_i & & & \\ & & \ddots & & \\ & & & & c_i \end{bmatrix} \tag{5}$$

and, combining equations (3) and (5):

$$\left[\frac{\partial X_i}{\partial Z_i}\right] = \left[-\frac{\partial F_i}{\partial X_i}\right]^{-1} \cdot \left[\frac{\partial F_i}{\partial Y_i}\right] \cdot \left[\frac{\partial Y_i}{\partial Z_i}\right]. \tag{6}$$

From equation (4),

$$\left[\frac{\partial Y_i}{\partial X_{i-1}}\right] = \begin{bmatrix} a_{i-1} & & & & \\ & a_{i-1} & & & \\ & & \ddots & & \\ & & & & a_{i-1} \end{bmatrix} \cdot \tag{7}$$

Thus, for discharges in the adjacent upstream reach,

$$\left[\frac{\partial X_i}{\partial Z_{i-1}}\right] = \left[-\frac{\partial F_i}{\partial X_i}\right]^{-1} \cdot \left[\frac{\partial F_i}{\partial Y_i}\right] \cdot \left[\frac{\partial Y_i}{\partial X_{i-1}}\right] \cdot \left[-\frac{\partial F_{i-1}}{\partial X_{i-1}}\right]^{-1} \cdot \left[\frac{\partial F_{i-1}}{\partial Y_{i-1}}\right] \cdot \left[\frac{\partial Y_{i-1}}{\partial Z_{i-1}}\right]. \tag{8}$$

Generalizing to more than two adjacent sections,

$$
\left[\frac{\delta X_i}{\delta Z_j}\right] = \begin{cases} \left[\frac{\delta X_i}{\delta Y_i}\right] \cdot \left[\frac{\delta Y_i}{\delta X_{i-1}}\right] \cdot \left[\frac{\delta X_{i-1}}{\delta Y_{i-1}}\right] \cdots \left[\frac{\delta X_{j+1}}{\delta Y_{j+1}}\right] \cdot \left[\frac{\delta Y_{j+1}}{\delta X_j}\right] \cdot \left[\frac{\delta X_j}{\delta Z_j}\right] & i > j \\[3mm] \left[\frac{\delta X_i}{\delta Z_i}\right] & i = j \quad (9) \\[3mm] 0 & i < j \end{cases}
$$

Calculations of the ecosystem response matrix by this method are presented graphically in Figure 6 for the non-dispersive case. A comment on these results will be held until the Jacobian method is presented.

Jacobian Method

The Jacobian method requires that the model be specified in the form

$$F = G(X) + CZ \tag{10}$$

where F is a vector of 242 (22 x 11) time derivatives, $G(X)$ are the rate expressions linking the endogenous variables, and CZ is a vector of discharges corrected to $mg\ell^{-1}day^{-1}$.

This much more straightforward representation allows the calculation of a 242 x 242 matrix, $\left[\frac{\delta X}{\delta Z}\right]$, in the same way as before, i.e.,

$$\left[\frac{\delta X}{\delta Z}\right] = \left[-\frac{\delta F}{\delta X}\right]^{-1} \cdot \left[\frac{\delta F}{\delta Z}\right] \tag{11}$$

but at great expense, since inverting this size matrix on an IBM 370/155 takes in the neighborhood of four minutes.

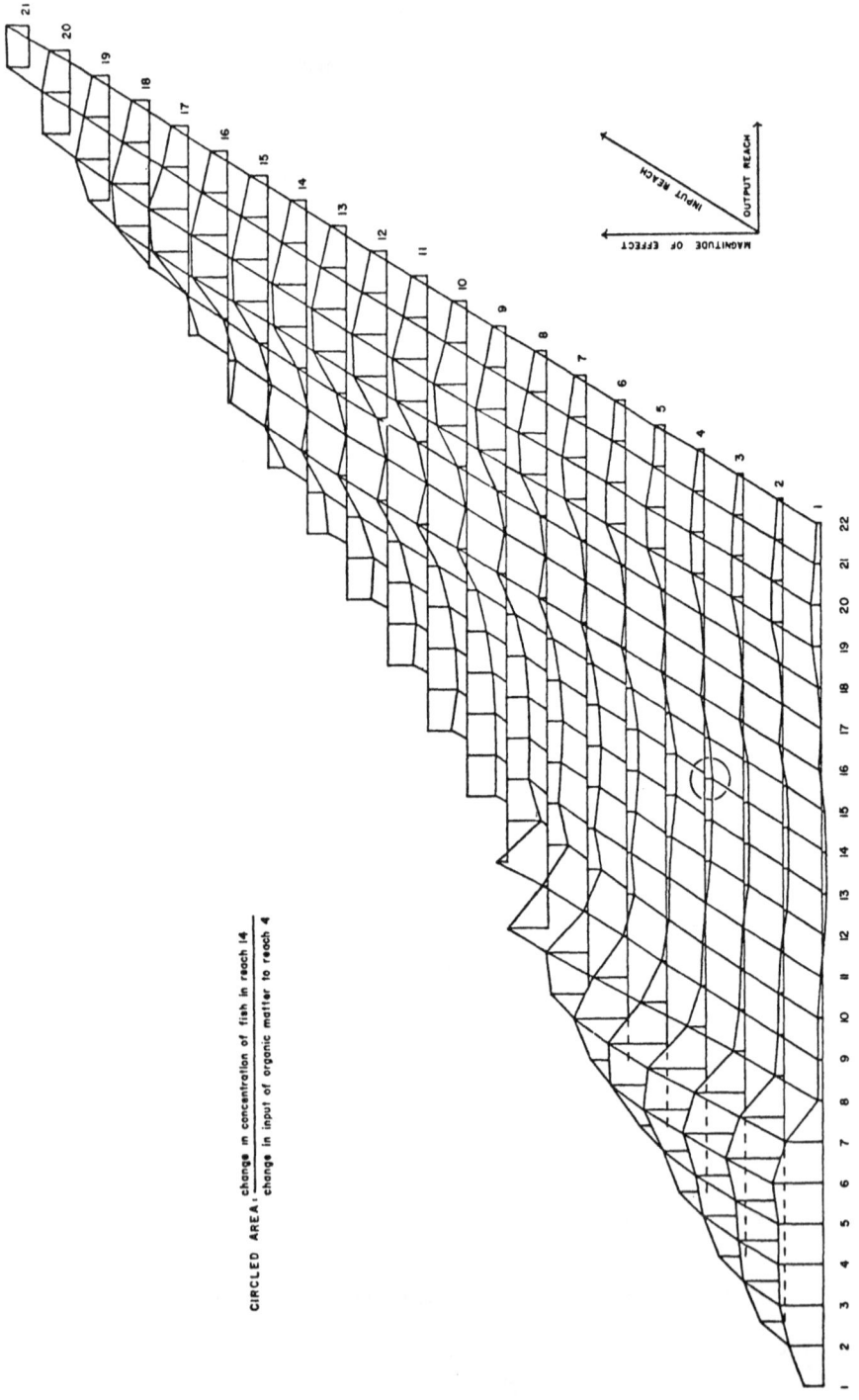

CIRCLED AREA: change in concentration of fish in reach 14 / change in input of organic matter to reach 4

Figure 6. Environmental response matrix for the change in fish with a change in input of BOD.

Comment

For the non-dispersive case, the results obtained by the two methods are identical (mathematically and numerically proven). However, the inclusion of dispersion requires that the Jacobian approach be used, adding significantly to computer time and core utilized (computer time is increased an order of magnitude, and core utilization is doubled). The greater accuracy of adding dispersion does not seem warranted at this time when the model is to be used in conjunction with the DVRMM, although the methodology and feasibility of including dispersion has been demonstrated.

V. Conclusions

It seems that two very broad conclusions can be reached in light of the discussion already given: first, ecological modeling, in a prediction sense, is still in its infancy. Much is known about the behavior of ecosystems and their responses to pollution, but little is known as to very specific and precise effects of changes in residuals inputs to a given system. In many cases it appears that even the direction of change of a particular component may be in question (see Figure 6). Second, very basic and simple models appear to be able to mimic ecological responses fairly well. In many cases, as I am sure will be shown at this conference, these models have been used as inputs to decision-making processes and thus can serve extremely useful functions, in spite of their crudity. I think their utility will prove even greater in the future.

References

Abbe, G. R. "An Evaluation of the Distribution of Fish Populations of the Delaware River Estuary." MS thesis (biological sciences), University of Delaware, 1967.

Anderson, G. C., and Zeutschel, R. P. "Release of Dissolved Organic Matter by Marine Phytoplankton in Coastal and Offshore Areas of the Northeast Pacific Ocean," Limnology & Oceanography, Vol. 15, no. 3 (1970), pp. 402-407.

Delaware River Basin Commission. Final Progress Report on Delaware Estuary and Bay Water Quality Sampling and Mathematical Modeling Project. Report to the Federal Water Pollution Control Administration, U.S. Department of the Interior. Trenton, N.J., DRBC, 1970.

Fair, G. W., and Geyer, J. C. Water Supply and Waste Water Disposal. New York: Wiley, 1956.

Herrman, R. B., Warren, C. E., and Doudoroff, P. "Influence of Oxygen Concentration on the Growth of Juvenile Coho Salmon," Transactions American Fisheries Society, Vol. 91, no. 2 (1962), pp. 155-167.

Ichthyological Associates. Ecological Study of the Delaware River in the Vicinity of Artificial Island. Trenton, N.J., IA, 1968.

_____. A Report on the American Shad and Other Anadromous Fishes Taken in Drifted Gill Nets in the Delaware River in the Vicinity of Artificial Island During March-June 1969. Trenton, N.J., IA, 1969.

_____. Ecological Study of the Delaware River in the Vicinity of Newbold Island. Trenton, N.J., IA, 1970.

Kelly, R. A. "Conceptual Ecological Model of the Delaware Estuary." To be published in B. C. Patten (Ed.), Systems Analysis and Simulation in Ecology, Vol. IV. New York: Academic Press, 1974.

_____, and Spofford, W. O., Jr. "Application of an Ecosystem Model to Water Quality Management: The Delaware Estuary." To be published in C. A. S. Hall and J. W. Day (Eds.), Models as Ecological Tools: Theory and Case Histories (New York: Wiley-Interscience, forthcoming).

McDonnell, A. J., and Hall, S. D. "Effect of Environmental Factors on Benthal Oxygen Uptake," Journal of Water Pollution Control Federation, Vol. 41, no. 8 (1969), pp. R353-R363.

Owens, M., and Maris, P. J. "Some Factors Affecting the Respiration of Some Aquatic Plants," Hydrobiologia, Vol. 23, no. 3-4 (1964), pp. 533-543.

Patten, B. C. "Mathematical Models of Plankton Production," Internationale Revue Gesamten Hydrobiologie, Vol. 53, no. 3 (1968), pp. 357-408.

_____. (Ed.) Systems Analysis and Simulation in Ecology, Vol. I. New York: Academic Press, 1971.

_____. (Ed.) Systems Analysis and Simulation in Ecology, Vol. II. New York: Academic Press, 1972.

Pielou, E. C. An Introduction to Mathematical Ecology. New York: Wiley-Interscience, 1969.

Russell, C. S., Kelly, R. A., and Spofford, W. O., Jr. "Early Returns on the Prospects for Regional Residuals Management Models." Paper presented at ORSA/TIMS meeting, Boston, Mass., April 1974.

_____, and Spofford, W. O., Jr. "A Quantitative Framework for Residuals Management Decisions," in A. V. Kneese and B. T. Bower (Eds.), Environmental Quality Analysis: Theory and Method in the Social Sciences. Baltimore: Johns Hopkins University Press, 1972.

Silver, S. J., Warren, C. E., and Doudoroff, P. "Dissolved Oxygen Requirements of Developing Steelhead Trout and Chinook Salmon Embryos at Different Water Velocities," Transactions American Fisheries Society, Vol 92, no. 4 (1963), pp. 327-343.

Smith, J. Models in Ecology. Cambridge, Mass.: The University Press, 1974.

Sokolnikoff, I. S., and Redheffer, R. M. Mathematics of Physics and Modern Engineering. New York: McGraw-Hill, 1958.

Spofford, W. O., Jr., Russell, C. S., and Kelly, R. A. "Operational Problems in Large Scale Residuals Management Models," paper presented at Universities-National Bureau of Economics Research Conference on Economics of the Environment, University of Chicago, 1972.

Thomann, R. V. Systems Analysis and Water Quality Management, Environmental Science Services Division, Environmental Research and Applications, Inc., New York, 1972.

U.S. Environmental Protection Agency. Estuarine Modeling: An Assessment, 16070 DZV 02/71, 1972.

_____. Delaware Estuary Water Quality Standards Study, August 1973.

U.S. Geological Survey. Water Resources Data for Pennsylvania, Part 2, Water Quality Records. Water Resources Division, U.S.G.S., P.O. Box 1107, Harrisburg, Pa. 17108. 1968.

Application of Mathematical Models to the Study,
Monitoring and Management of the North Sea

Jacques C. J. Nihoul

1. THE NORTH SEA MODELLING EFFORT

Mathematical models of the North Sea have been developed
in most of the bordering countries. The national efforts are
now coordinated by the Joint North Sea Modelling Group initia-
ted by JONSIS (the Joint North Sea Information System) repor-
ting to I.C.E.S. (the International Council for Exploration of
the Sea). Data are provided by international surveys called
"JONSDAP" (Joint North Sea Data Acquisition Program) (Fig. 1).

In an earlier stage, the models tended to address separately the Physics, the Chemistry and the Biology of the North
Sea. Now, the development of computing facilities allowing
more ambitious programs, these models are progressively integrated in a common, general, interdisciplinary model with the
purpose of understanding the North Sea environment, predicting
its evolution, - taking into account the constraints of Modern
Society -, and assisting its management.

The distinction, between complex research models and simple,
oriented management models, is avoided. Management models are
regarded as "subsets" of the general multipurpose model, derived
from it to answer specific questions with the degree of sophistication which the objectives, on the one hand, the reliability
of data, on the other hand, recommend.

The different stages in the elaboration of the mathematical
model and of its submodels are shown in fig. 2.

1. The mathematical description of the system is confronted
with the data base constituted from existing or newly acquired
data (e.g. Jonsdap campaigns, Belgian five years survey of the
Eastern part of the Southern Bight,...). The data base provides

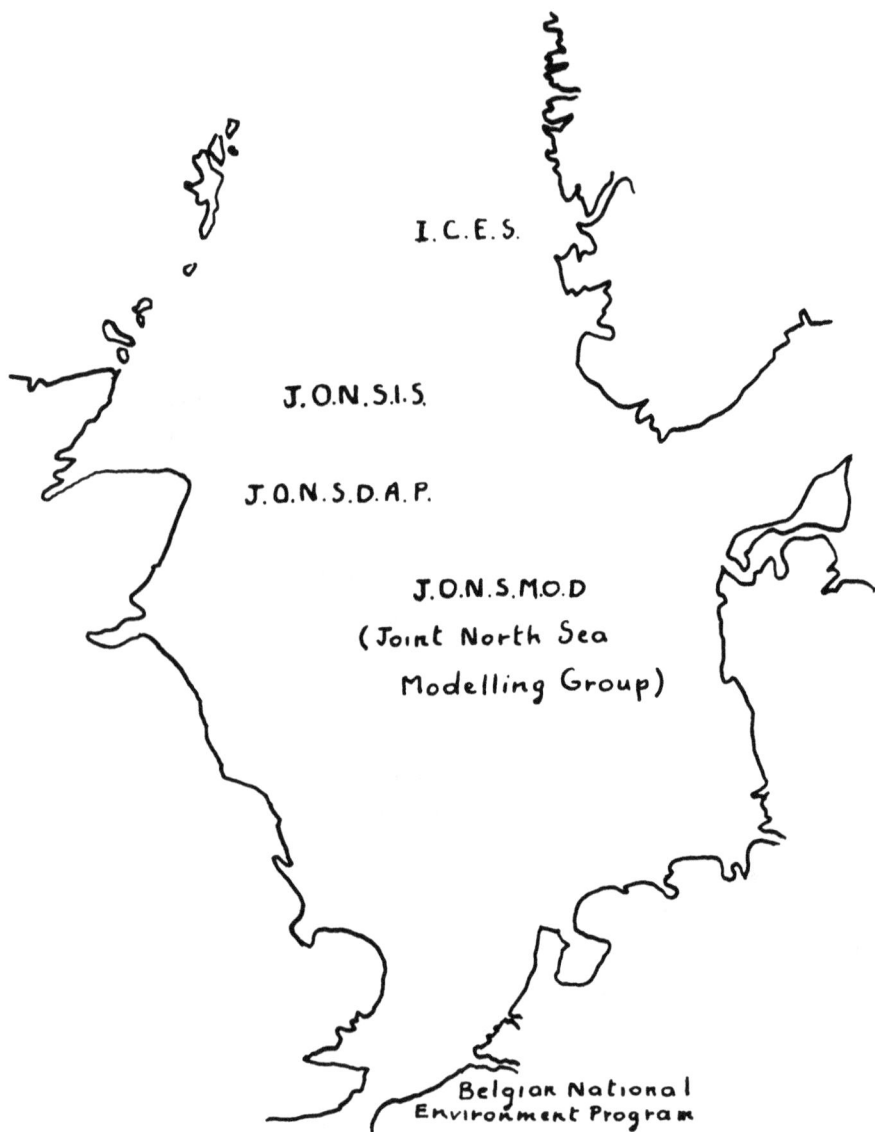

I.C.E.S.

J.O.N.S.I.S.

J.O.N.S.D.A.P.

J.O.N.S.M.O.D
(Joint North Sea
Modelling Group)

Belgian National
Environment Program

Figure 1.   Southern bight of the North Sea

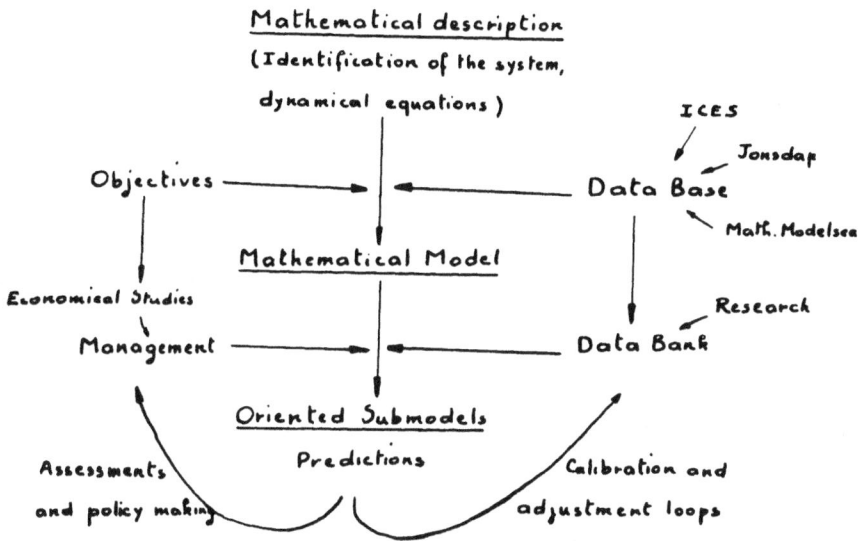

Figure 2. Elaboration of the mathematical model

(i)   a correlation study suggesting variables which are not
      significantly interrelated and between which interactions
      may be disregarded,

(ii) an orders of magnitude study indicating variables and pro-
      cesses which can be neglected,

(iii) a sensitivity analysis evaluating the degree of refine-
      ment which is required in the specification of state va-
      riables and interaction laws,

(iv) a dialogue with the users of the model allowing a more
      precise definition of the objectives and indicating the
      degree of sophistication which is required to produce re-
      liable predictions answering the questions put to the
      model without unnecessary expensive complexity.

2.   The mathematical model which emerges from the confrontation
with the data base is submitted to the scientific reflection.

      Data processing, simulation tests and fundamental research
contribute to a better understanding of the structure of the
system and of the ability of the model to describe it. The
assessment of the modelling prospective efficiency combined with
specific management requests determines subsets of the general
model which can be used for reliable, speedy predictions answering limited purposes and assisting immediate decisions.

      The calibration, adjustment and exploitation of the sub-
models feed back information in the general model whose development, combined with new requests from management objectives,
gives birth to second generation submodels with increased reliability.

      A continuous interaction between research and application
is thus achieved in the guiding framework of mathematical modelling.

2. THE MODELS OF THE NORTH SEA

To proceed from a general mathematical description to a tractable mathematical model and later to more limited sub-models one can (Nihoul 1974) :

1) reduce the *support*, i.e. the extent of the system in physical space and time either by (i) narrowing the field of investigation or (ii) averaging over one or several space coordinates or over time,

2) reduce the *scope*, i.e. the dimensions of the system in state space either by (i) closing the system at a limited number of state variables (allowing for the global effect of other less essential parameters in adjustable coefficients) or (ii) averaging over suitably defined compartments of which only the aggregate properties, not the details, are described.

The operating models of the North Sea can all be regarded as reduced size versions of an interdisciplinary three-dimensional model on which one or several simplifications were performed as described above.

i) Integration over depth and reasonable hypotheses on the vertical density distribution have allowed the development of two dimensional models of tides and storm surges.

ii) Further time integration (over a time sufficiently long to cover several tidal periods and thus cancel to a large extent tidal oscillations and transitory wind currents) has given the residual circulation model where the results of tidal computations are used to calculate the forcing due to non-linear tidal interactions.

iii) Tidal and residual models have been exploited to evaluate
the dispersion and the advection of marine constituents
and to elaborate dispersion models adapted to the study
of coastal discharges of pollutants and off-shore dumpings.

iv) The hydrodynamic models have revealed distinctive marine
regions where different current regimes prevail and which
appear as natural boxes for the elaboration of completely
space integrated time dependent chemical and ecological
box models.

Chemical and ecological models have also been simplified
by restricting attention to the aggregate properties of compartments such as dissolved substances, suspensions, bottom sediments,
phytoplankton, zooplankton, heterotrophic bacteria, fish[*],...

Figs 3 and 4 (Math. Modelsea 1974) illustrate the quality of
the results obtained with the operating tides and storm surges
models. They show the lines of equal amplitudes and equal phases
drawn from the observations and calculated by the model. An
excellent agreement is found.

Fig. 5 (Ronday 1975) shows the residual circulation in the
North Sea computed by the model, using the predictions of the
tidal model to calculate the forcing term produced by non-linear
tidal oscillations. One essential result of this model (Math.
Modelsea 1974) is the revelation of residual gyres (regions of
closed stream lines) which earlier models had failed to uncover.

---

[*] In this approximation, if one is interested, for instance, in
cadmium pollution, one takes as state variables of the model
the total amounts of cadmium in solution, suspension, in
phytoplankton ... and writes the dynamic evolution equations
expressing the "translocations" of cadmium from one compartment to the other.

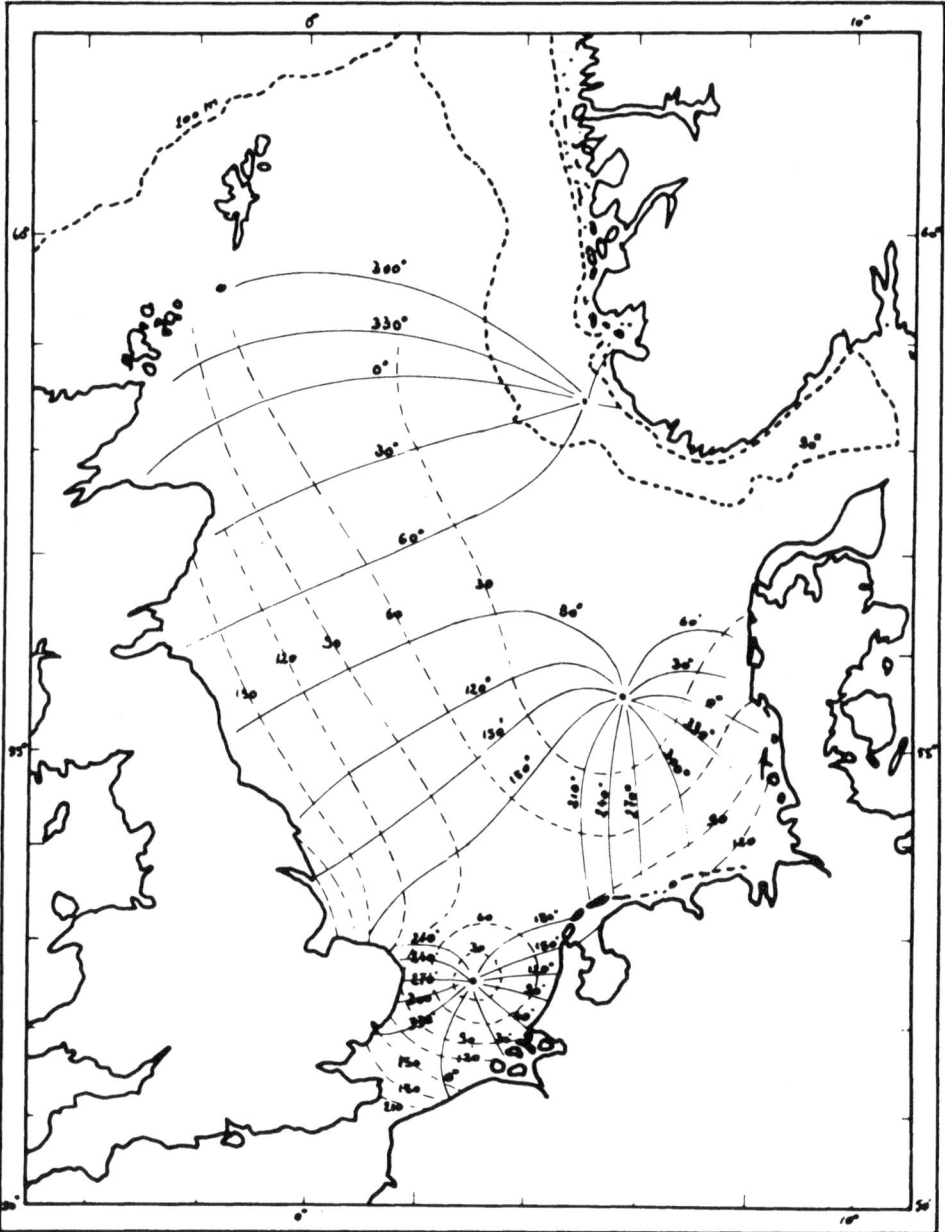

Figure 3. Results from tidal model

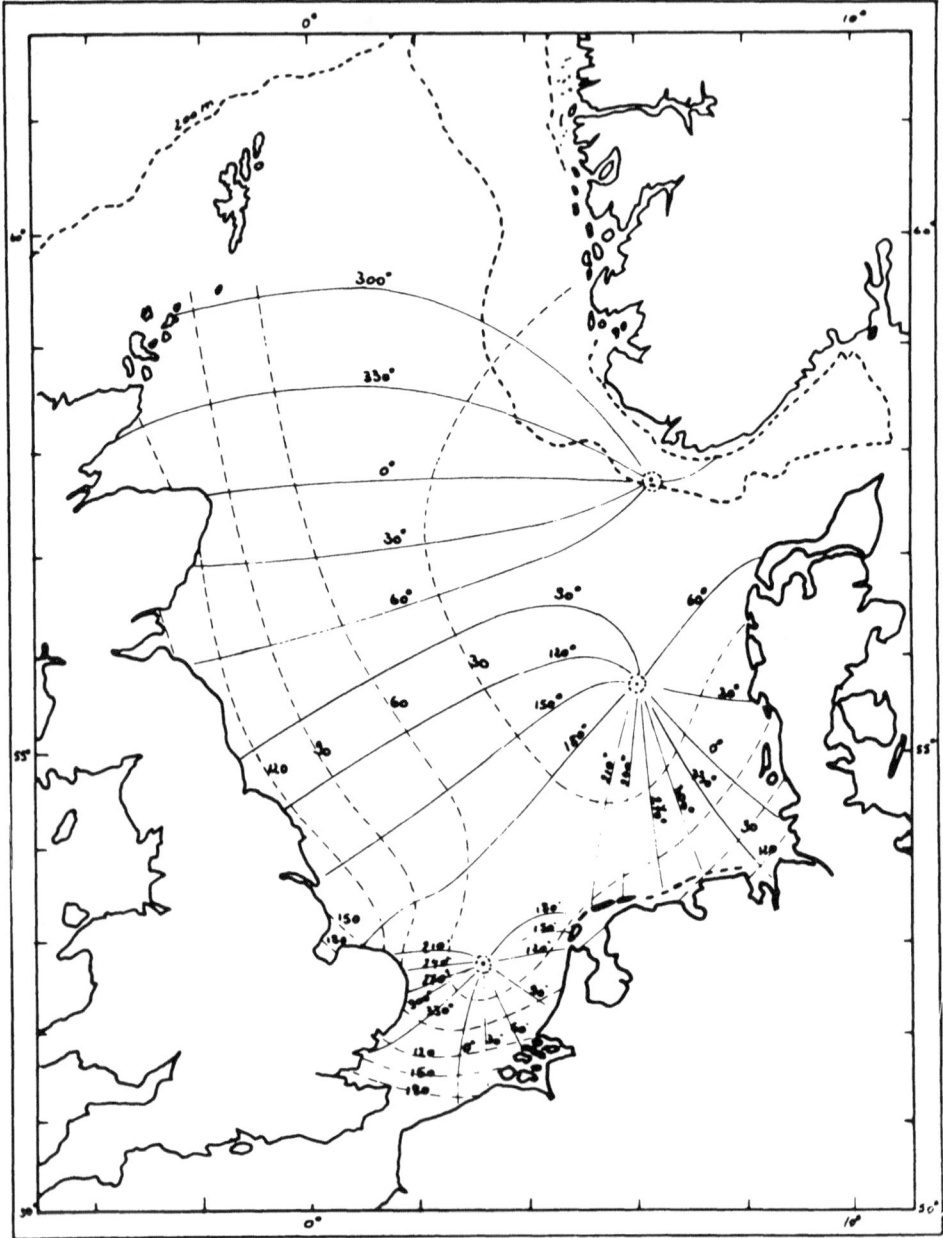

Figure 4. Results from storm surge model

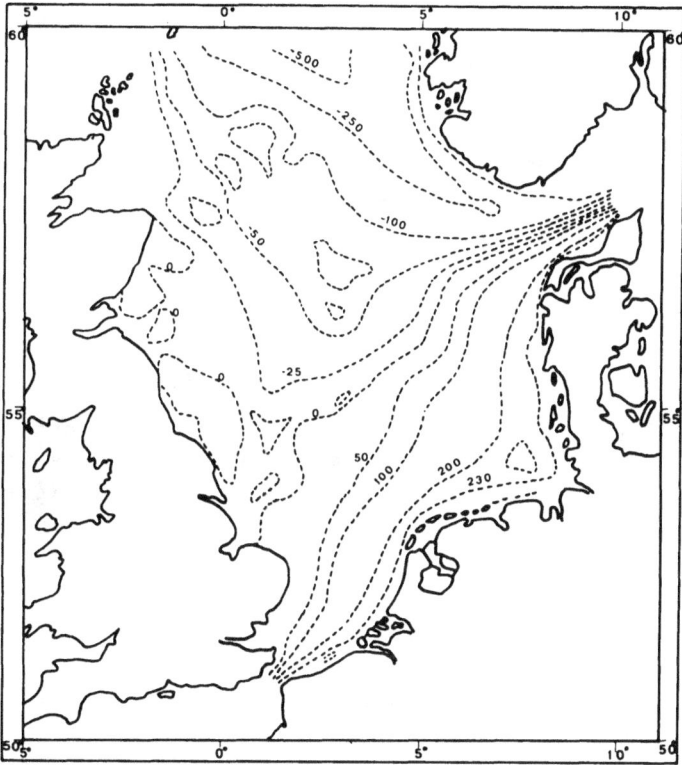

Figure 5. Computed residual circulation from the North Sea

3. APPLICATION TO THE SOUTHERN BIGHT

The management requirements for the Southern Bight of the North Sea concern essentially :

i)   the dispersion of turbidity and pollutants around the Scheldt estuary, the existing or potential dumping grounds and the coastal outfalls,

ii)   the appraisal of the health and stability of coastal eco-systems subject to pollution stresses such as those descri-bed in (i),

iii) the understanding of sedimentation and bottom erosion with particular emphasis on mud accumulation for coastal engi-neering problems.

The model demonstrated that in this shallow area of extreme-ly variable depth, the dispersion of pollutants was dominated by the shear effect associated with the vertical variations of very intense tidal currents. The predictions of the size and the shape of patches of pollutants after a release were found in excellent agreement with the observations (Nihoul 1974).

The presence of residual gyres was identified as a major factor in the sedimentation pattern and the existence of ecologi-cal niches where distinct ecosystems prevail.

In particular, the gyre which can be seen off the eastern Belgian coast on fig. 5[**] succeeded in explaining the observed accumulation of mud and heavy metals in the bottom sediments along the coast by the entrainment and prolonged residence of highly turbid waters from the Scheldt estuary.

[**]   It is partly masked in the whole North Sea pattern by the coarseness of the grid but it appears in a clear undeniable way in the results of the Southern Bight model using a finer grid (Nihoul 1975).

The model showed that the gyre created, in that region, outer-
lagoon conditions characterized by high nutrient concentrations
and phytoplankton biomass but little zooplankton grazing and
intensive recycling of nutrients by bacteria (revealing a rather
unhealthy short-circuited food chain where additional releases
of nutrients might create the conditions of eutrophication)
(Nihoul 1975).

A nutrient cycle box model was elaborated with special emphasis on the coastal gyre region. On that basis, models were
derived to simulate the translocation of pollutants (such as
heavy metals) from the water column, through the food chain, to
the consumable fish. The predicted concentrations in fish were
found in good agreement with the measured concentrations in
sampled specimens (Math. Modelsea 1974).

4.   ASSISTANCE TO MANAGEMENT

The Southern Bight model can predict with great accuracy the
elevation of the water surface produced by tides and storm surges,
the dispersion, sedimentation - and eventual recirculation by
strong turbulence - of suspended material as well as the final
deposition of sediments on the bottom. It can simulate the effect
of coastal engineering works (dredging, construction of a harbour
...) and can give fully assistance to management in this respect.

By revealing the existence of distinctive regions where different circulation regimes prevail, the model identified ecological
niches which are the natural boxes for adjacent box models describing the dynamics of the Bight's ecosystems. These models can
evaluate the fluxes of carbon, nitrogen,...., pollutants... through
the food chain and provide an estimate of the anticipated fish
population and level of pollution. In this respect the model can

assist Public Health decision. Equivalently, it can elaborate on Public Health tolerances to determine acceptable upper bounds for the pollutants'concentrations in coastal waters or sediments.

The model can predict the dispersion pattern of pollutants both in the water column and in the sediments. By evaluating the extent of the damage produced by a given coastal or off-shore release, the model can thus appreciate the opportunity of authorizing or penalizing dumpings in the sea and assist management decision. Furthermore, determining the transfer functions which relate the intensity of the source (the amount released) and the final concentrations in the sea, the model, working backwards from Public Health tolerances, can set up for management the problem of optimizing, subject to economical constraints, the tolerable inputs and the locations of sources of pollution, coastal outfalls and sea dumpings.

5. REFERENCES

Math. Modelsea (1974), Mathematical models of Continental Seas,
Dynamic Processes in the Southern Bight, I.C.E.S. Hydrography
Committee C.M. 1974-C : 1 .

Nihoul, J.C.J. (1974), Modelling of Marine Systems, Elsevier
Publ. Co, Amsterdam, 264 p.

Nihoul, J.C.J. (1975), Mesoscale secondary flows and the dynamics
of ecosystems in the Southern Bight of the North Sea.
Mémoires Soc. Sc. Lg., to be published. Proc. 6th Liège Coll.
on Ocean Hydrodynamics, 1974.

Ronday, F.C. (1975), Mesoscale effects of the tidal stress on the
residual circulation of the North Sea, Mémoires Soc. Sc. Lg.,
to be published. Proc. 6th Liège Coll. on Ocean Hydrodynamics,
1974.

PHYTOPLANKTON MODELS AND EUTROPHICATION PROBLEMS

Donald J. O'Connor, Dominic M. Di Toro and Robert V. Thomann

Environmental Engineering and Science Program
Manhattan College, New York City

The primary purpose of this paper is to present the applications of a set of equations describing the seasonal distribution of phytoplankton to the analysis of eutrophication problems in various locations throughout the country. A brief review of the theoretical structure of the analysis is presented in Part I with a qualitative description of the pertinent equations and a discussion of the general procedure of the verification process. Examples from various natural water systems are presented in Part II to demonstrate the utility of this type of analysis in evaluating alternate plans to restore or maintain appropriate levels of water quality. The systems considered are: the fresh water segment of the San Joaquin River (1); the estuarine regions of the Sacramento-San Joaquin Delta (2,3); and the Potomac River (4); Western Lake Erie (5,6); and Lake Ontario (7). The individual studies cited here contain complete bibliographies of the scientific and engineering literature which was used in the analysis of the general problem and the development of the equations, and since these rather extensive references are recorded in the above papers and reports, they are not repeated in this paper.

I. THEORETICAL STRUCTURE OF THE ANALYSIS

The general principle of conservation of mass is the basis which is used to formulate equations of the various constituents of importance in the analysis of the eutrophication problem, the essential element of which is the dynamic behavior of phytoplankton. In their simplest form, these equations relate the growth of phytoplankton to the availability of nutrients; in their most complex form they may incorporate the interaction among many nutrients, specification of a number of species of phytoplankton, and predator-prey relationships between contiguous trophic levels. The nutrients, which may be in various stages of reaction themselves, are supplied not only by inputs from man-made and natural sources, but also by recycling from internal sources due to death and decay of the organic matter in the system. These equations are expressed most fundamentally in terms of the rates of change of the interactive substances. In this form, they also provide the greatest insight and understanding of the phenomenon. These differential equations are developed by applying the principle of conservation of mass from which quantitative relationships are developed relating to the progressive changes of state of the various constituents in time and space. Furthermore, solutions for equilibrium or steady-state conditions may be available which follow directly from the differential equations or their integrated forms.

The basic constituents of the analysis are the nutrients, phytoplankton and the zooplankton, for each of which an equation is developed as follows: Consider a segment of a natural water system of specified volume. The mass rate of change of the constituent within it is the product of the volume V and change of concentration, Δc, over the time interval Δt, and is accounted for by the rates of change of three components: the physical transport, J, through the volume, V; the chemical and biological transformations within it, R; and the inputs to or withdrawals from it, W. The mass rate of change for any segment is written:

$$V \frac{dc}{dt} = J \pm \sum R \pm \sum W \qquad (1)$$

Although the latter two terms are usually combined as sources and sinks, they are maintained as separate components in this paper. This distinction is made in order to emphasize the practical importance of the inputs, over which there exists the maximum possible control from the viewpoint of water quality management.

TRANSPORT STRUCTURE

The transport component is due to the advective and dispersive characteristics of the hydrodynamic regime and the settling characteristics of the various constituents. In many rivers and estuaries the primary transport is one dimensional along the longitudinal axis of flow, while in most lakes the significant transport is in the vertical plane. In the former, the settling component may not be significant. The mixing and turbulence, associated with both fresh water flow and tidal action, are frequently of sufficient magnitude to minimize the settling effect. In lakes, however, vertical mixing is usually less intense and settling is invariably a significant factor. In addition to longitudinal and vertical transport, lateral dispersion may also play an important role, particularly between the shallow littoral or embayment areas and the deep central segments of the water body.

The transport patterns which apply to the examples in this paper are shown in Figure 1. A purely advective system, as shown in Figure 1A, is the simplest transport model. Each segment, which is assumed to be of uniform concentration, receives input from the adjacent upstream segment and discharges to the downstream, without dispersive exchange between contiguous segments. The only transport factor is the hydraulic flow. The transport coefficient is the reciprocal of the detention time in the segment. The San Joaquin model is comprised of one such segment. Figure 1B presents an advective-dispersive pattern, typical of estuaries and some fresh water rivers. In addition to the advective component, transport also includes dispersion due to tidal circulation. The figure also indicates the lateral embayments which exchange with the main body. This pattern represents the transport structure of both the Sacramento-San Joaquin Delta and the Potomac Estuary. An expanded lateral dimension is shown in Figure 1C which is representative of large bays, lakes and oceans. Advection and dispersion may occur through any of the interfaces shown. Western Lake Erie is characterized by such a model.

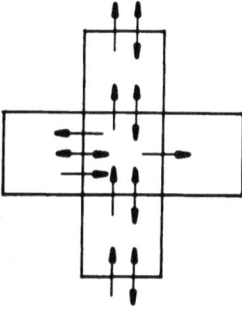
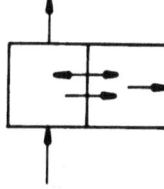
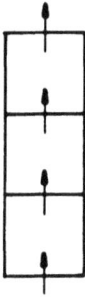
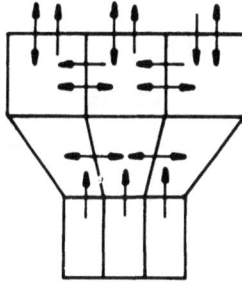
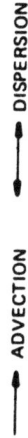

Fig. 1. Transport structure.

The above represents the typical horizontal patterns of transport in fresh water streams, tidal rivers, estuaries, lakes and
littoral ocean regions. In addition, the vertical component may
have a significant effect due to such factors as upwelling, settling
or diffusion from benthal deposits. Figure 1D shows a simplified
vertical profile representing the transport pattern used for the
analyses of Lake Ontario.

KINETIC INTERACTIONS

The kinetic elements of the equations incorporate two broad
categories: the biological growth and death of the phytoplankton,
zooplankton , and upper levels of the food chain, and the biochemical
transformations of the nutrients, nitrogen and phosphorus. These
regimes are linked by the growth coefficient of the phytoplankton,
which is a function of temperature, light and nutrient concentrations.
The respiration, death and excretion of the various trophic levels
produce organic forms of nitrogen and phosphorus, some of which may
settle. These, in turn, break down to yield inorganic nutrients
which are available to the phytoplankton for growth. Ammonia
is also subjected to biochemical oxidation by bacteria. These
interactions are shown diagramatically in Figure 2.

1.   Phytoplankton

The growth rate of phytoplankton is a function of temperature,
light and nutrient concentration. The growth coefficient is directly
related to temperature in moderate climates. It is also dependent

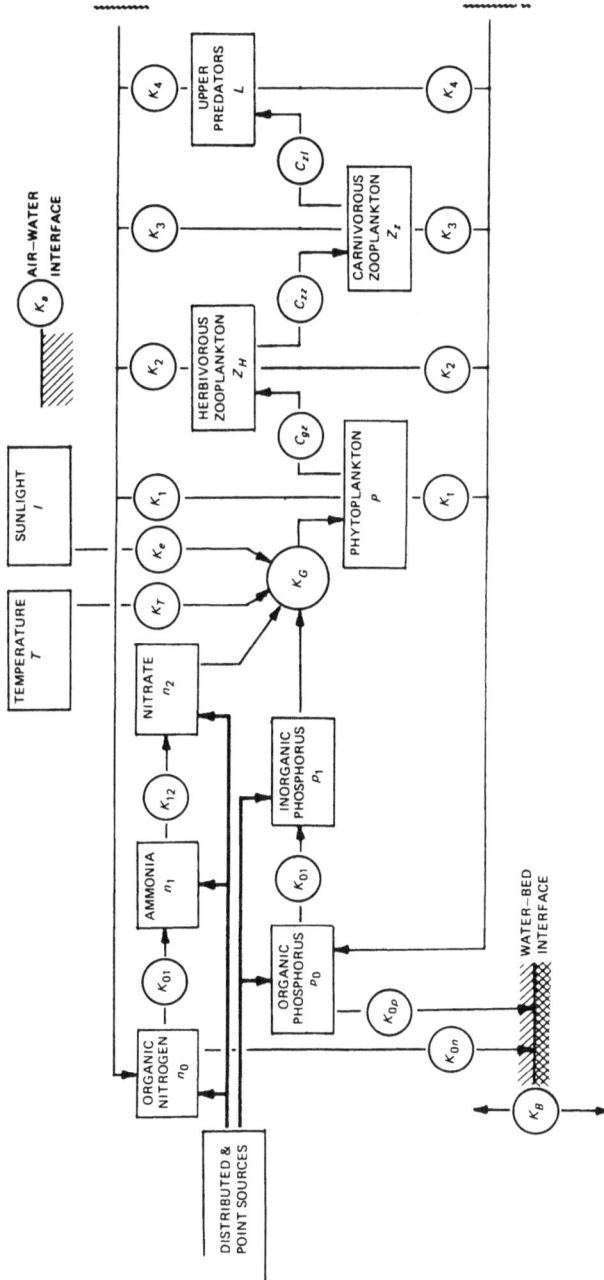

Fig. 2. Kinetic interrelationships.

on the light intensity and nutrient concentrations up to a saturating
or limiting condition, greater than which it decreases with light
and remains constant with nutrient concentration. The latter is
described by a Michaelis-Menton formulation whose significant parameter is that concentration at which the growth rate is equal to
one-half of that at the saturated concentration. If more than one
nutrient is involved, the growth rate is assumed to be proportional
to the product of the Michaelis expressions for each of the nutrients.
A preference structure of ammonia over nitrate is built into the
analysis.

The effect of non-optimal light conditions is to reduce the
growth rate. Furthermore, in a natural environment, the available
light decreases with depth. The extinction coefficient describes
the exponential decay. In addition, the surface light varies throughout the day. The mean daily value for the photoperiod fraction of
the day is used in these analyses. The light effect on growth is,
therefore, time averaged over the day and vertically averaged over
the depth.

The removal of phytoplankton is caused by endogeneous respiration, a function of temperature, and is also due to predation by
the zooplankton concentrations. The pertinent parameter in the
latter is the grazing coefficient.

The kinetic equation which embodies these reactions is

$$R_p = (G_P - D_P)\ P \tag{2}$$

G_p is the growth rate expression:

$$G_p = K_T(T) \; r(I_s, K_e) \; \frac{n}{K_{Mn} + n} \; \frac{p}{K_{Mp} + p} \tag{3}$$

made up of the saturated growth rate K_T (T), which is a function
of temperature, T; the reduction due to non-optimal incident
light r, a function of the saturating light intensity, I_s, and the
extinction coefficient, K_E; and the nutrient reduction factors,
for inorganic nitrogen, n, and orthophosphorus, p, with Michaelis
constants K_{Mn} and K_{Mp}. The death rate expression is:

$$D_p = K_1(T) + C_g(T) \mathcal{Z} \; \frac{K_{MP}}{K_{MP} + P} - \frac{w}{H} \tag{4}$$

made up of the endogenous respiration rate $K_1(T)$, the grazing rate
$C_g(T)$ the zooplankton biomass $\mathcal{Z}$, the Michaelis constant for zooplankton grazing K_{MP}, the settling velocity, w and the depth H.

2. Zooplankton

This system is analogous to that of the phytoplankton. Zooplankton grow in accordance with the availability of their food, the
phytoplankton. The zooplankton, in turn, are predated upon by the
upper levels of the food chain and undergo endogeneous respiration
and death. Their excretion products may be significant sources of
nutrients. Furthermore, to account for the fact that the zooplankton
graze more than they consume, a conversion efficiency coefficient is
introduced into these equations. In order to simplify the analysis,
the predation effect of the upper levels is usually introduced as
an empirical constant. The kinetic equation is:

$$R_Z = (G_Z - D_Z) \, Z \qquad (5)$$

$$G_Z = a_{ZP} \, C_g(T)P \, \frac{K_{MP}}{K_{MP} + P} \, a_1 \qquad (6)$$

Where a_1 is the assimilation efficiency, a_{ZP} is the carbon to chloro-phyll ratio and C_g and K_{MP} are as defined previously. D_Z is the death rate expression.

$$D_Z = K_2 \, (T) \qquad (7)$$

where $K_2(T)$ is the endogenous respiration coefficient.

In the analysis of Lake Ontario, it was necessary to include predation by the next trophic level. Therefore, the zooplankton are divided into two classes: the herbivorous, which prey on the phytoplankton, and the carnivorous, which in turn prey on the herbivorous.

3. Nitrogen

The major components of the nitrogen system are detrital organic nitrogen, ammonia nitrogen, and nitrate nitrogen. In natural waters there is a stepwise transformation from organic nitrogen to ammonia, nitrite and nitrate, yielding nutrients for phytoplankton growth. The kinetics of these transformations are assumed to be first order reactions with temperature-dependent rate coefficients. The hydrolysis of organic nitrogen yields ammonia, which is oxidized through nitrite to nitrate.

Two sources of detrital organic nitrogen are considered: organic nitrogen produced by the endogenous respiration of phytoplankton and zooplankton, assuming only organic forms of nitrogen result from this process; and the organic nitrogen equivalent of grazed but not metabolized phytoplankton excreted by zooplankton. Analyses of mass balances of nutrients indicate that, in some cases, substantial loss of material occurs. It is hypothesized that this loss is due to settling of the particulate fraction of total nitrogen. To incorporate this effect into a depth-averaged formulation a fraction of the organic nitrogen is removed, presumably by settling.

The equation for the reaction term of organic nitrogen, n_O is:

$$R_{nO} = a_{nP} D_P P + a_{nP} a_{ZP} D_Z Z + a_{nP} P \left[1 - \frac{a_1 a_{ZP} K_{MP}}{K_{MP} + P} \right] C_g Z \qquad (8)$$

$$- K_{On1} n_O - K_{On} n_O$$

where a_{nP} is nitrogen to chlorophyll ratio, a_{ZP} is the carbon to chlorophyll ratio, K_{On1} is the decay of organic nitrogen to ammonia and K_{On} is any additional decay or removal coefficient.

The first and second terms represent organic nitrogen released through endogenous respiration by the phytoplankton and zooplankton respectively. The third term represents the organic nitrogen of the grazed but unassimilated phytoplankton. The last terms represent the decomposition, settling and other effects that contribute to the overall removal of organic nitrogen.

For ammonia, n_1, and nitrate, n_2, the reaction terms are:

$$R_{n1} = K_{On1} n_O - K_{12} n_1 - a_{nP} \alpha G_P P \qquad (9)$$

$$R_{n2} = K_{12} n_1 - a_{nP} [1 - \alpha] G_P P \qquad (10)$$

where K_{On1} is the rate of production of ammonia from organic nitrogen and K_{12} is the rate of oxidation of ammonia to nitrate. If the ammonia is preferentially assimilated by phytoplankton, a preference coefficient, α, is introduced, which specifies that ammonia is used until its concentration reaches the range of the inorganic nitrogen half-saturation constant, at which point the nitrogen source shifts to nitrate.

4.   Phosphorus

The phosphorus system is similar in some respects to that of nitrogen. Organic phosphorus is generated by the death of phytoplankton. Organic phosphorus, as in the case of the organic nitrogen, represents the nonliving detrital material. Phosphorus in this form is then converted to the inorganic state, where it is available to the algae. A sink of organic phosphorus, like nitrogen, is due to settling, which may be significant depending on the magnitude of vertical mixing. If settling is effective, it is assumed that only the organic form of the nutrients is susceptible to removal in this fashion.

$$R_{po} = a_{pP} D_P P + a_{pP} a_{ZP} D_Z Z + a_{pP} P \left[1 - \frac{a_1 a_{ZP} K_{MP}}{K_{MP} + P} \right] C_g Z \quad (11)$$

$$- K_{op1} P_o - K_{op} P_o$$

where $a_p P$ is the phosphorus to chlorophyll ratio , K_{op1}, is the decay to inorganic phosphorus and K_{op} represents any removal or additional decay of organic forms. For the inorganic phosphorus, p_1, which is available to the phytoplankton, the reaction equation is:

$$R_{p1} = K_{op1} P_o - a_{pP} G_p P - K_{1p} p_1 \quad (12)$$

where K_{opl} represents the rate of decomposition of organic phosphorus to inorganic forms available for phytoplankton utilization and K_{lp} represents any additional loss of inorganic phosphorus.

The kinetic interactions of these elements are shown in Figure 2: primary production, which converts inorganic nutrients to phytoplankton; secondary production of zooplankton, accomplished by their grazing on phytoplankton; mortality and excretion pathways, which release organic material in particulate and soluble form; the deposition pathway, which accounts for whatever settling of particulate organic material occurs; and the regeneration pathways, which convert organic forms into inorganic forms that are then available for primary production.

INPUTS

The sources of nutrients are the runoff from natural drainage and urban developments and discharges from municipal, industrial and agricultural activities. These inputs, which are both point and distributed, are included in the various constituent equations in both the organic and inorganic forms. Data on point sources such as tributary rivers and direct municipal and industrial discharges are usually available. Those sources are known, identifiable and controllable. However, information on distributed sources is frequently lacking, and estimates must be made in such cases. Inputs from these sources are more difficult to quantify, and control measures are usually much more expensive. Therefore, great emphasis is being placed on distinguishing between the two from the water quality management point of view.

The sources described above are external to the system, by contrast to the internal sources, such as the nutrients recycled in the water column or released from the bed in biological and bio-chemical transformations. Mass transfer rates from these sources may be of the same order of magnitude as the external sources and, in some cases, many times greater. The relative significance of the external and internal sources is of paramount importance in evaluating the feasibility of various control measures.

The inflow hydrograph, solar radiation, photoperiod and extinc-tion coefficient are the remaining external variables to specify. Depending on the importance of these factors, independent models may be used to describe the spatial and temporal distribution of the relevant parameter. The output from such models supplies the input to the eutrophication analysis. Alternatively, empirical fits to the distribution of these external parameters is usually sufficient to describe their effect.

SEGMENTATION OF THE SYSTEM

The model is thus composed of a set of simultaneous nonlinear, partial differential equations in time and space. In order to solve these equations, it is invariably necessary to resort to finite difference techniques in either time or space or both. For the latter dimension, the water body is divided into a series of segments, such that the assumption of spatial homogeneity is maintained without violating certain mathematical requirements or introducing unnecessary computational complexity. The three basic factors, transport, reactions and inputs should be taken into account in this process. The segmentation of the system should allow for a realistic portrayal of the transport phenomenon for both the advective and dispersive components. It should permit a reasonable representation of the kinetics, particularly with respect to the shallow littoral zones, which are more productive than the deep central sections of the water body. Finally, it should provide a finer definition in the vicinity of the inputs, where relatively steep gradients exist by comparison to the more distant locations where the effects are less pronounced. Invariably, the size and number of segments is a compromise between these criteria and the capacity of computational facilities available. The number of components and segments is ultimately limited by the latter consideration.

II. APPLICATIONS OF THE MODEL

The model described in the previous section has been applied, with appropriate modifications, to the analysis of phytoplankton distributions in various water systems throughout the country. The examples presented in this section are abstracted from studies on the freshwater stretch of the San Joaquin River, the estuarine segments of the Sacramento-San Joaquin Delta, the Potomac Estuary, the western basin of Lake Erie and Lake Ontario.

The general format is as follows: a brief description of the area and the problem is presented. The specific structure of the

model within the general framework presented in Figure 2 is discussed. Transport, kinetics and inputs are specified in each case.
The calibration and validation procedure is briefly described and
comparisons between the model output and observed data are presented.
Applications of the model to the particular water quality problem
in each area are shown.

A summary of the physical characteristics of the systems,
the transport coefficients, the kinetic parameters and inputs are
presented in Table 1. The transport and kinetic coefficients vary,
of course, in accordance with the hydrological and climatological
conditions of the areas. It is significant that the kinetic parameters fall within the range of those values reported in the literature
and furthermore fall in a relatively narrow band within this range.
The consistency of these coefficients adds further substantiation
to the validity of the approach.

SACRAMENTO-SAN JOAQUIN DELTA

The specific question which was addressed in this study is the
potential eutrophication problem in the delta and downstream channels
and bays which may arise if freshwater from the Sacramento River is
diverted for use in other areas in California. The reduced freshwater flow has two effects: an increase in the detention or flushing
time of the system, and an increase in light transmission due to a
possible decrease in the input of suspended sediment. The analysis
also includes the effect of the project increases in nutrient discharges from the municipal, industrial and agricultural development

Table 1

PARAMETER	UNITS	SYMBOL	RIVER SAN JOAQUIN	ESTUARIES DELTA	ESTUARIES POTOMAC	LAKES ERIE	LAKES ONTARIO
PHYSICAL							
Temperature Range	°C	T	8-25	7-26	5-30	2-25	1-20
Solar Radiation Range	Langleys/day	I	200-700	160-720	135-530	170-780	100-630
Detention Time	days	t_o	30	30	60	45	7.9 yrs
Extinction Coefficients	1/m	K_e	4.0	5.5	3.5	1.2	0.22
PHYTOPLANKTON							
Saturating Light Intensity	Langleys/day	I_s	300	300	300	350	350
Saturated Growth Rate	1/day@20°C	K_r	2.0	2.5	2.0	1.3	2.1
Engogenous Respiration Rate	1/day@20°C	K_l	0.10	0.10	0.10	0.08	0.10
Nitrogen Michaelis Constant	μg/1	K_{mn}	25.0	25.0	25.0	25.0	25.0
Phosphorus Michaelis Constant	μg/1	K_{Mp}	-	-	5.0	10	2.0
Settling Velocity	m/day	W	-	-	-	-	0.1
ZOOPLANKTON							
Grazing Rate	ℓ/mg carbon-day@20°C	C_g	0.13	0.18	-	0.25	1.2
Endogenous Respiration Rate	1 day@20°C	K_2	0.075	0.10	-	0.16	0.02
Assimilation Efficiency	mgC/mgC	a_l	0.60	0.60	-	0.65	0.60
Grazing Michaelis Constant	ug chlor/ℓ	K_{MP}	50	50	-	50	10
STOICHIOMETRIC RATIOS							
Carbon/Chlorophyll	mg/mg	a_{zP}	50	50	50	50	50
Nitrogen/Chlorophyll	mg/mg	a_{NP}	7	7	10	7	10
Phosphorus/Chlorophyll	mg/mg	a_{pP}	-	-	1	1	1

Table 1

| PARAMETER | SYMBOL | UNITS | RIVER | ESTUARIES | | LAKES | |
			SAN JOAQUIN	DELTA	POTOMAC	ERIE	ONTARIO
BIOCHEMICAL							
Org. Nitrogen Decomposition Rate	K_{On1}	1/day@20°C	-	-	0.14	0.04	0.035
Org. Nitrogen Removal Rate	K_{On}	1/day@20°C	-	-	0.10	0.028	0.001
Nitrification Rate	K_{12}	1/day@20°C	-	-	0.20	0.052	0.04
Org. Phosphorus Decomposition Rate	K_{op1}	1/day@20°C	-	-	0.14	0.40	0.14
Org. Phosphorus Removal Rate	K_{op}	1/day@20°C	-	-	0.10	0.028	0.001
Inorganic Phosphorus Removal Rate	K_{1p}	1/day@20°C	-	-	0.10	0	0

in the area.

The portion of the Delta being considered, as illustrated
in Figure 3A. is both tidal and saline. The transport coefficients
were established by the freshwater discharge and mean tidal flow
through the various segments, as shown in this figure: Benicia,
Suisun Bay, Honker Bay, Grizzly Bay, Chipps Island, lower Sherman
Island, Antioch and Big Break. In addition, the Suisun and Antioch
segments were subdivided into relatively shallow, shoreline areas
and a deeper central zone. There are a total of ten segments, the
length of each being on the order of a tidal excursion length or less.
A schematic of the system is shown in Figure 3B, indicating the
transport vectors for each element.

The monthly average flow used in the analysis is shown in Figure
4A; and the monthly average temperature, sunlight and photoperiod in
Figure 4B together with the functional representation of each used in
the computation. The data points are monthly averages based on daily
measurements.

The detention time of each segment is the volume divided by the
freshwater flow, and the tidal exchange is approximately given by the
tidal prism divided by the mean tidal volume of each segment. The
former coefficient varies temporally with the freshwater flows.
Given the flow as input data and the volumes of the segments, as specified, the detention time in each element is readily computed, and
varies over the year inversely as the patterns shown in Figure 4A.
The tidal exchanges between segments, on the other hand, are held

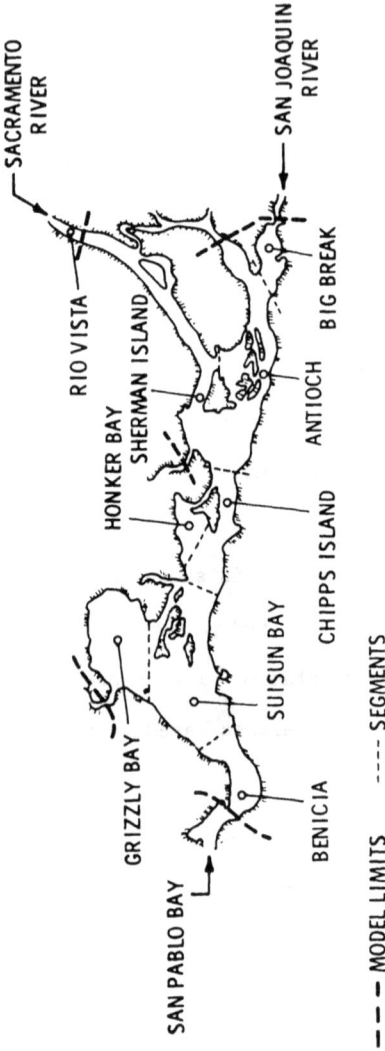

Fig. 3A. location map of study area.

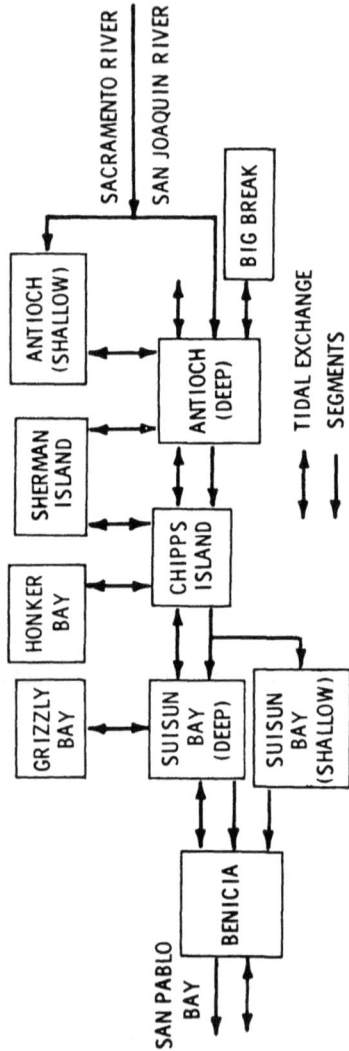

- - - MODEL LIMITS   ---- SEGMENTS

Fig. 3B. Segmentation and transport.

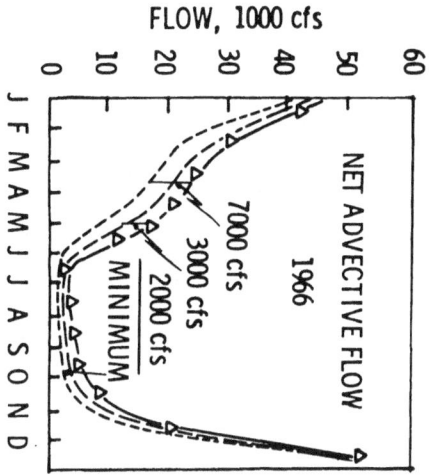

FIG. 4A. FLOW DISTRIBUTION

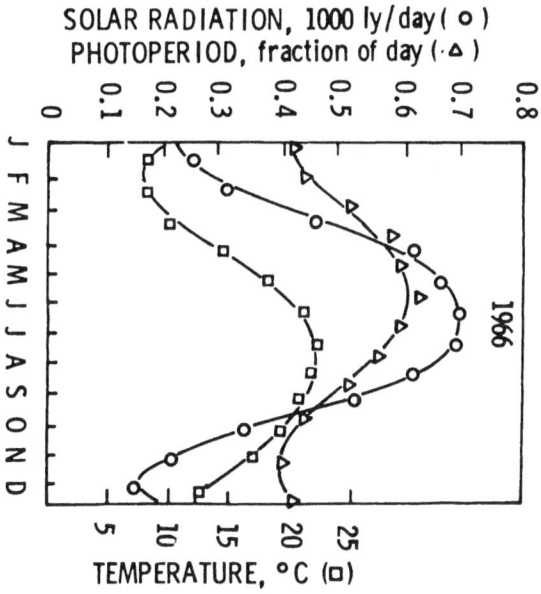

FIG. 4B. TEMPERATURE, LIGHT, AND PHOTOPERIOD.

constant over the period of analyses and calculated as noted above. The spatial distribution of the chloride concentration was used to check the transport coefficients. The nutrient inputs to the system are due to municipal, industrial and agricultural discharges.

The major limiting kinetic pathway of this analysis is the nitrogen cycle. Since phosphorus was present at all times of the year throughout the region of concern in concentrations greatly in excess of the Michaelis half-saturation constant, it was assumed that this element did not limit the growth of the phytoplankton. Therefore, the phosphorus cycle was not incorporated in the analysis. Furthermore, measurements of ammonia consistently indicated very low concentrations of this nutrient. It was also concluded that the bacterial nitrification pathway was not controlling and therefore was not included. Nutrient regeneration from excretion and death of the plankton were recycled directly to the nitrate form. The simplified model therefore consisted of three steps: inorganic nitrate conversion to phytoplankton, thence to zooplankton, with regeneration from both levels back to nitrate. Subsequent analysis incorporated the nitrification step, recognizing the potential importance of ammonia discharges due to increased population and industrial development. Because of the intensity of mixing due to tidal action, settling was assumed to be insignificant. The remaining elements of the analysis as described previously were incorporated.

Figure 5 is a comparison of the calculated distributions and the observations at Antioch for phytoplankton chlorophyll, zooplankton

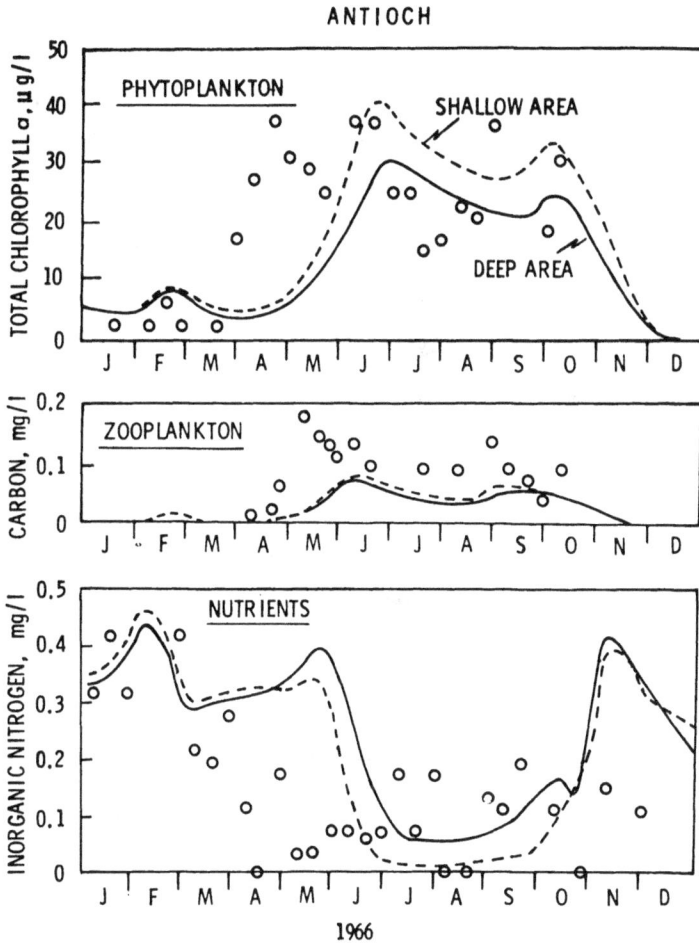

Fig. 5. Observed and computed profiles.

carbon, and total inorganic nitrogen. Differences are evident, but
the general pattern of the observations are reproduced by the model.
The calculated concentration of chlorophyll during the summer period
is in the range of 30-40 μg/l which is in substantial agreement with
the observed data. The calculated concentration of nitrogen also
agrees reasonably with the observations, considering the precision
of the chemical technique of measurement.

Figures 6A and B present comparisons of the calculated and observed
chlorophyll concentration for the channel areas (Antioch and Suisun)
and the bay regions (Grizzly and Suisun Bays). The calculated summer
conditions average approximately 30 μg/l in Suisun Bay, which com-
pares favorably with the observed values. The comparison of the
nutrient data at these stations is similar to that at Antioch. The
coefficients, which were used to calculate the above distributions,
were adjusted to the appropriate temperature, sunlight and flow conditions in 1967 to verify the model.

It is informative to compare the seasonal distribution at these
locations, all of which are subject to tidal action, to that at a
station at which the transport is solely due to freshwater flow.
Figure 7 presents the input variations for temperature, flow, solar
radiation, and the observed and calculated output at Mossdale, which is
further upstream in the freshwater section of the San Joaquin River.
It is interesting to note the more pronounced secondary bloom in
the latter part of each year and the significance of the freshwater
flow over the two-year period. The higher runoff in 1967 essentially

1966

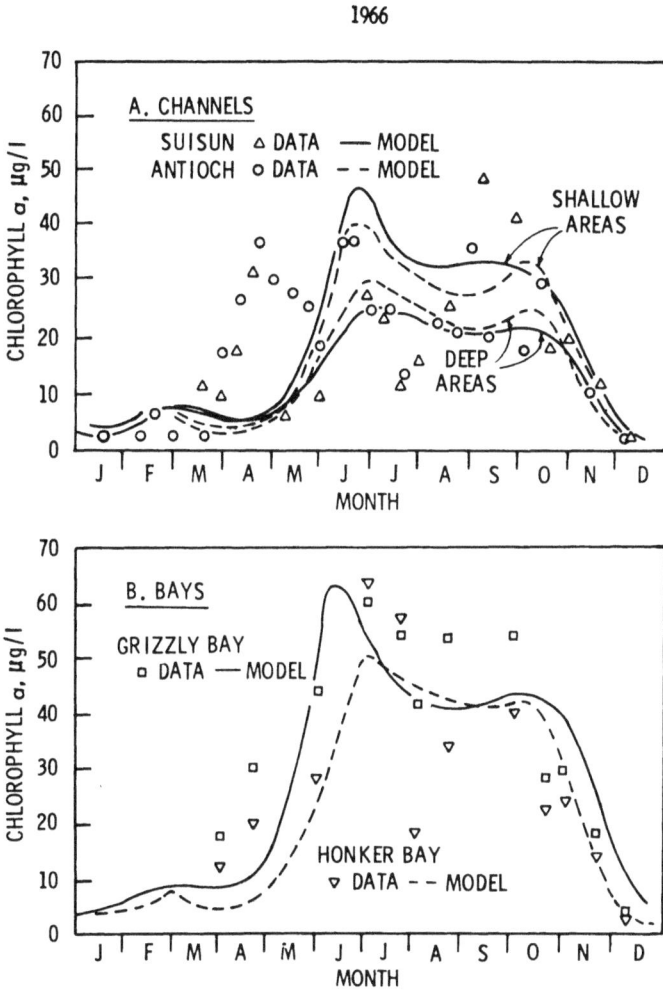

Fig. 6. Observed and computed profiles.

174

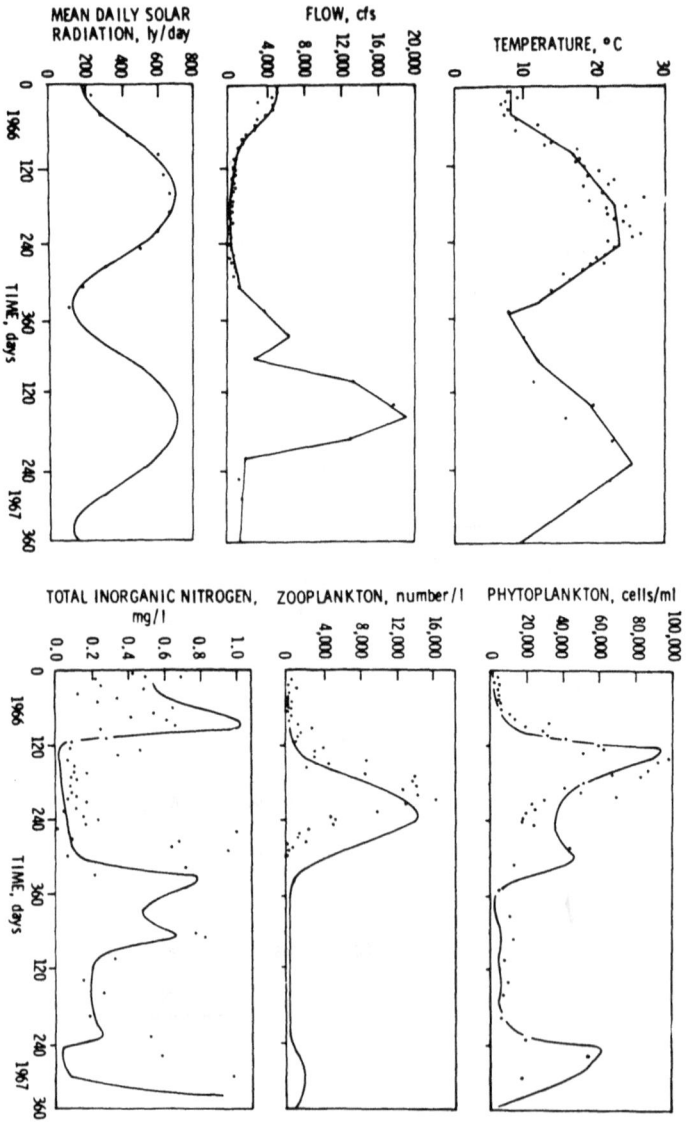

Fig. 7. San Joaquin River, Mossdale.

flushed out the system before the phytoplankton had an opportunity
to grow, in contrast to conditions in 1966 at a period of significantly
lower flow.

The proposed diversion plan would result in a reduced flow
through the system as well as a possibility of increased light transmission due to a reduction in suspended sediment concentration. In
addition, the municipal, industrial and agricultural growth in the
area will increase the nutrient input to the system. These inputs
are based on the population and economic growth of the area as projected in a recent study of the area.

1.   Increased Nutrient Input

The nitrogen inputs employed in this calculation are those projected for 1980, while all other conditions, the hydrograph, extinction coefficients and the background nitrogen level, were those
observed in 1966. The estimated 1980 nitrogen input is approximately
159,000 pounds/day. The profiles in Figure 8A indicate maximum
chlorophyll concentration at Antioch of about 90 ug/l, as compared
to concentrations observed in 1966, on the order of 40 μg/l. Projections and data are also presented for Suisun Bay where maximum
concentrations of approximately 150 μg/l of algal chlorophyll are
to be anticipated for the 1980 conditions.

2.   Diversion of Freshwater Flow

Figure 8B indicates the influence of the reduced freshwater
flow on chlorophyll concentration for the projected 1980 nitrogen

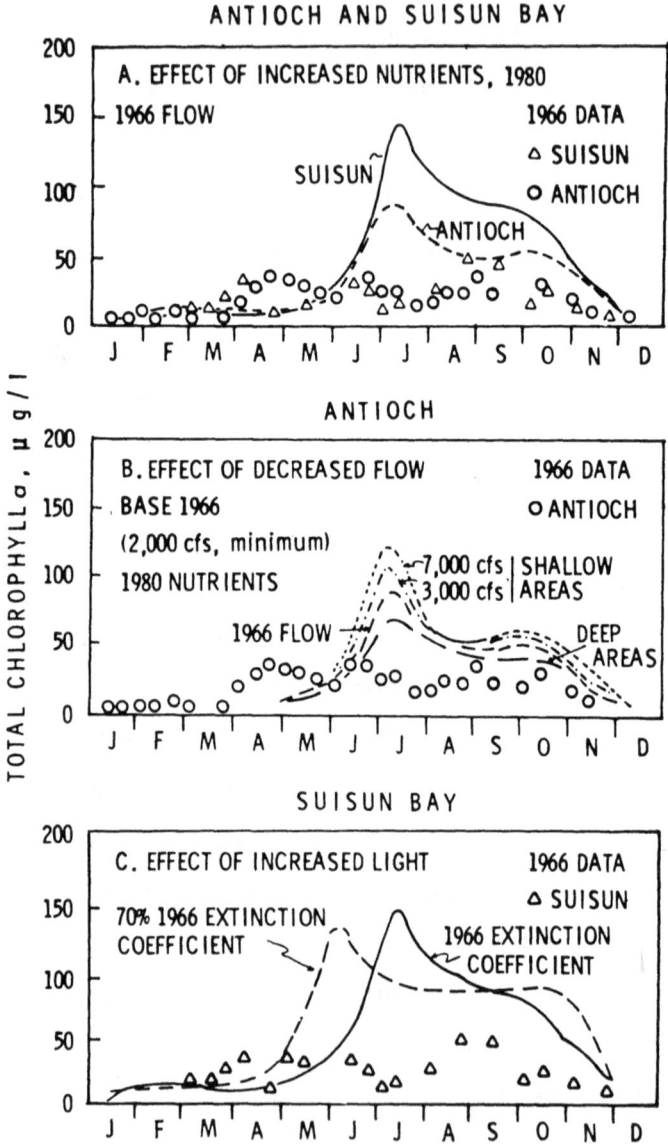

Fig. 8. Projections.

inputs. A possible range of flow diversions maintaining a minimum flow of 2,000 cfs, is shown in Figure 4A. These projections were developed for 1966 extinction coefficients, background nitrogen, phytoplankton and zooplankton levels. The 1966 hydrograph with 1980 inputs yields a peak phytoplankton concentration of approximately 90 ug/l as indicated above. Reducing the flow by 3,000 and 7,000 cfs yields phytoplankton concentrations of approximately 100 µg/l and 120 µg/l, respectively. Similar projections for Grizzly and Suisun Bays provide basically the same indications concerning the effect of flow on the projected level of phytoplankton.

3.   Extinction Coefficient

In order to evaluate the effect of a possible reduction in turbidity and a consequent increase in light transmission, several analyses were conducted, one of which is presented in Figure 8C for Suisun Bay. The 1980 nitrogen discharge projections with 1966 flows were employed to calculate the profiles with the extinction coefficient varying as indicated. The extinction coefficients for the solid curve were those measured in 1966. The chlorophyll concentration reaches approximately 150 µg/l. The dashed curve presents the distribution for a reduction in the extinction coefficients to 70% of their value in 1966. For this condition, the chlorophyll concentration rises to a peak value which approximates the previous concentration and then levels off at about 100 µg/l for summer and early fall periods. However, the spring bloom occurs earlier. Further reductions in extinction coefficients produce more marked changes in both the channels and bays.

4.   Evaluation of Control Measures

In view of the magnitude of the concentrations of phytoplankton
which may be anticipated in the future, various alternates to control the phytoplankton levels were evaluated. The removal of nitrogen from the wastewater discharges was specifically considered,
assuming 80% as a technologically and economically feasible level
of removal. Figure 9 presents a summary of the projects at Antioch.
Figure 9A indicates the effect of flow diversions of 3,000 and 7,000
cfs with the 1966 extinction coefficients, while 9B presents the
effect of varying the extinction coefficients with the 1966 flows.
Figure 9C indicates the effect of providing 80% removal by treatment
of the anticipated 1980 waste discharges for the 1966 flow and
various extinction coefficients. The resulting effluents would discharge approximately 30,000 pounds, which is close to the 1966 condition. Since the projected levels of phytoplankton are equal to or
less than present concentrations, present available technology would
permit flow diversion in moderate quantities.

POTOMAC RIVER ESTUARY

The water quality of the Potomac Estuary has been the subject of
investigation over many decades. The area of interest, which is the
upper 40-mile reach of the Potomac Estuary, is generally not subject
to the intrusion of salts from Chesapeake Bay. In the vicinity of
Washington, D.C., the river is tidal with an average depth of 18 feet
and is characterized by numerous coves and embayments whose depths

ANTIOCH

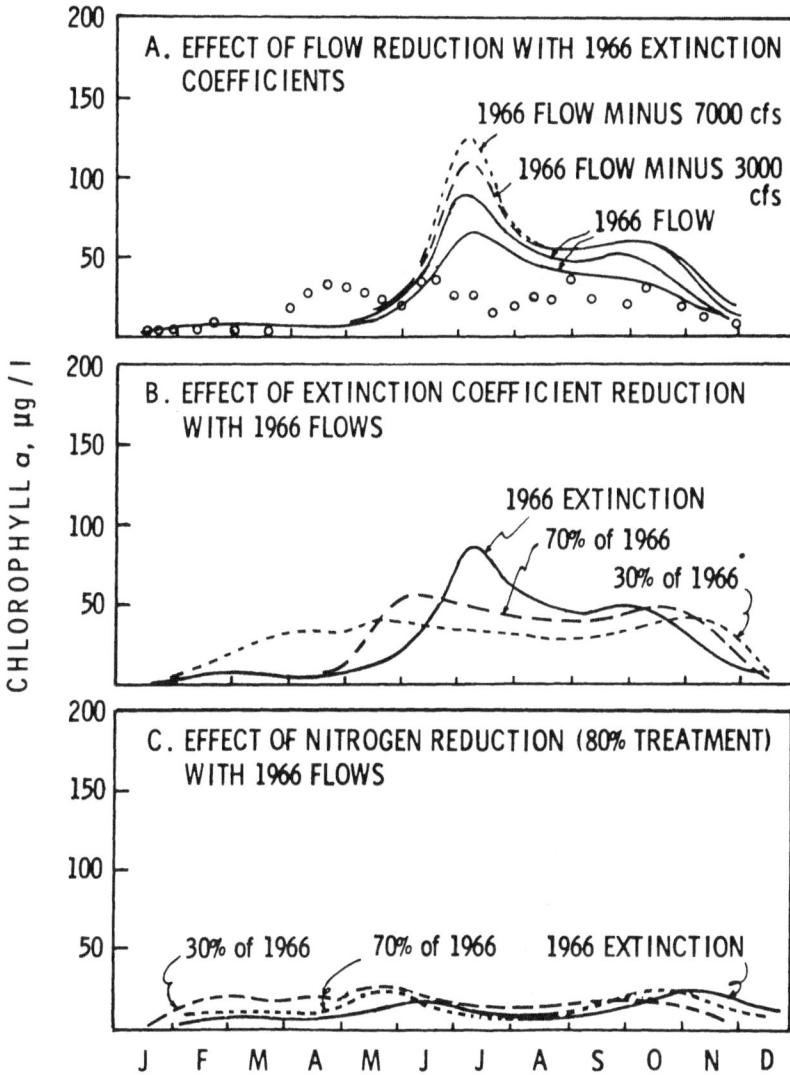

Fig. 9. Effect of flow reduction, extinction reduction, and
nitrogen input reduction.

are significantly less than that of the main estuary. Currently,
the major waste input is the effluent from the Washington, D.C.,
secondary treatment plant. Spatially, the estuary is divided into
23 longitudinal segments as shown in Figure 10. An additional 15
spatial segments were incorporated into the model to reflect the
effects of the shallow tidal flats. A total of 38 spatial segments
is used to represent the Potomac with primary emphasis on the Upper
Estuary in the vicinity of Washington, D.C.

The model incorporates interactions among the following variables
of Figure 2: phytoplankton, organic nitrogen, ammonia nitrogen,
nitrate nitrogen, organic phosphorus, and inorganic phosphorus.
Several runs were made to examine the sensitivity of constituents to
various levels of zooplankton. It was concluded that the predatory
effect was minimal, at least for the upper end of the estuary downstream from Washington, and therefore the effect was not included
in the analysis. This is supported by the fact that the phytoplankton of concern in this area is Anacystis, a blue-green alga that
is toxic to zooplankton.

The calibration and verification of the model was accomplished
using the data sets of 1968 and 1969. Comparisons between observations and calculations are shown in Figure 11 and 12. In general,
the model as formulated provides a reasonable approximation to
1969 conditions. Recognizing that the calculated profiles are based
on the coefficients obtained from the analysis of the previous
year's data, a basis is therefore established for the credibility
of the model. Spatial profiles during July 1969, compare favorably

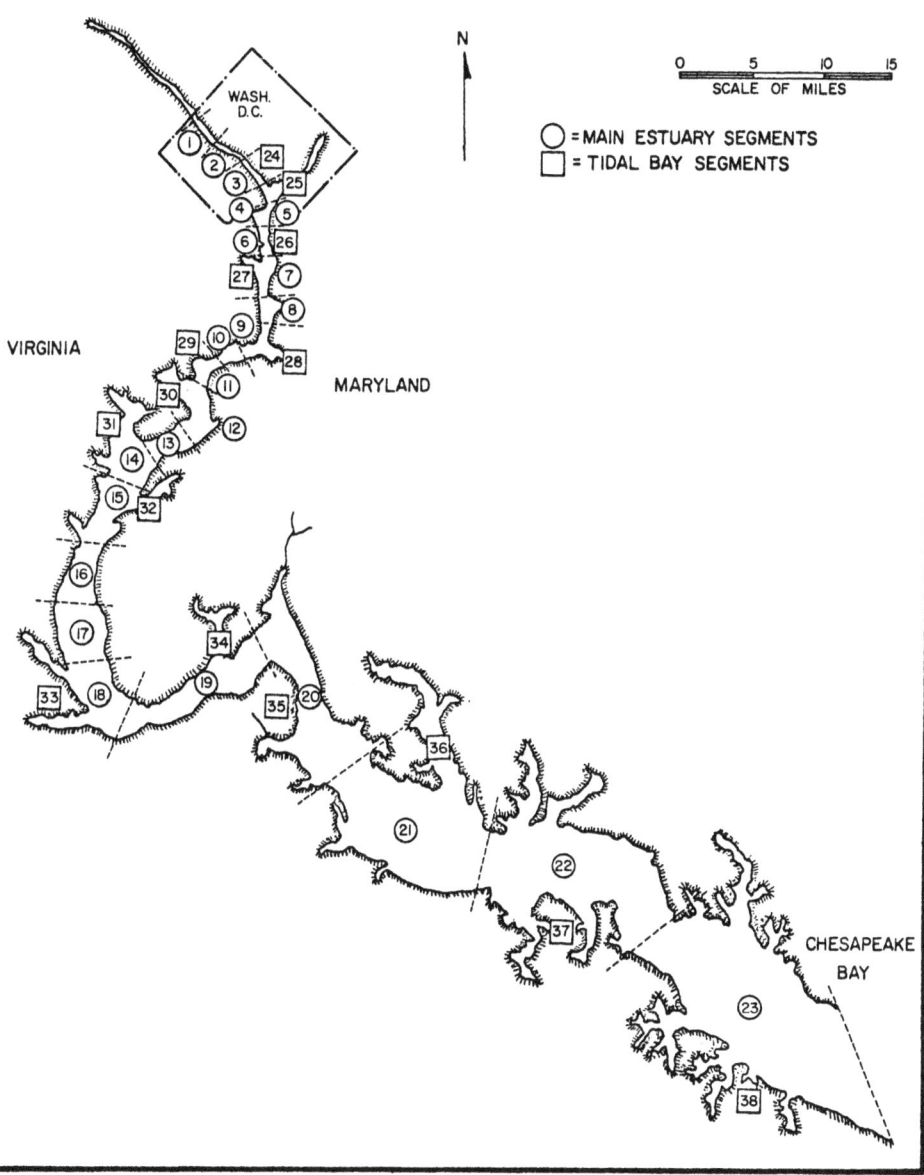

FIG. 10.-MAP OF POTOMAC ESTUARY SHOWING LONGITUDINAL AND LATERAL SEGMENTS.

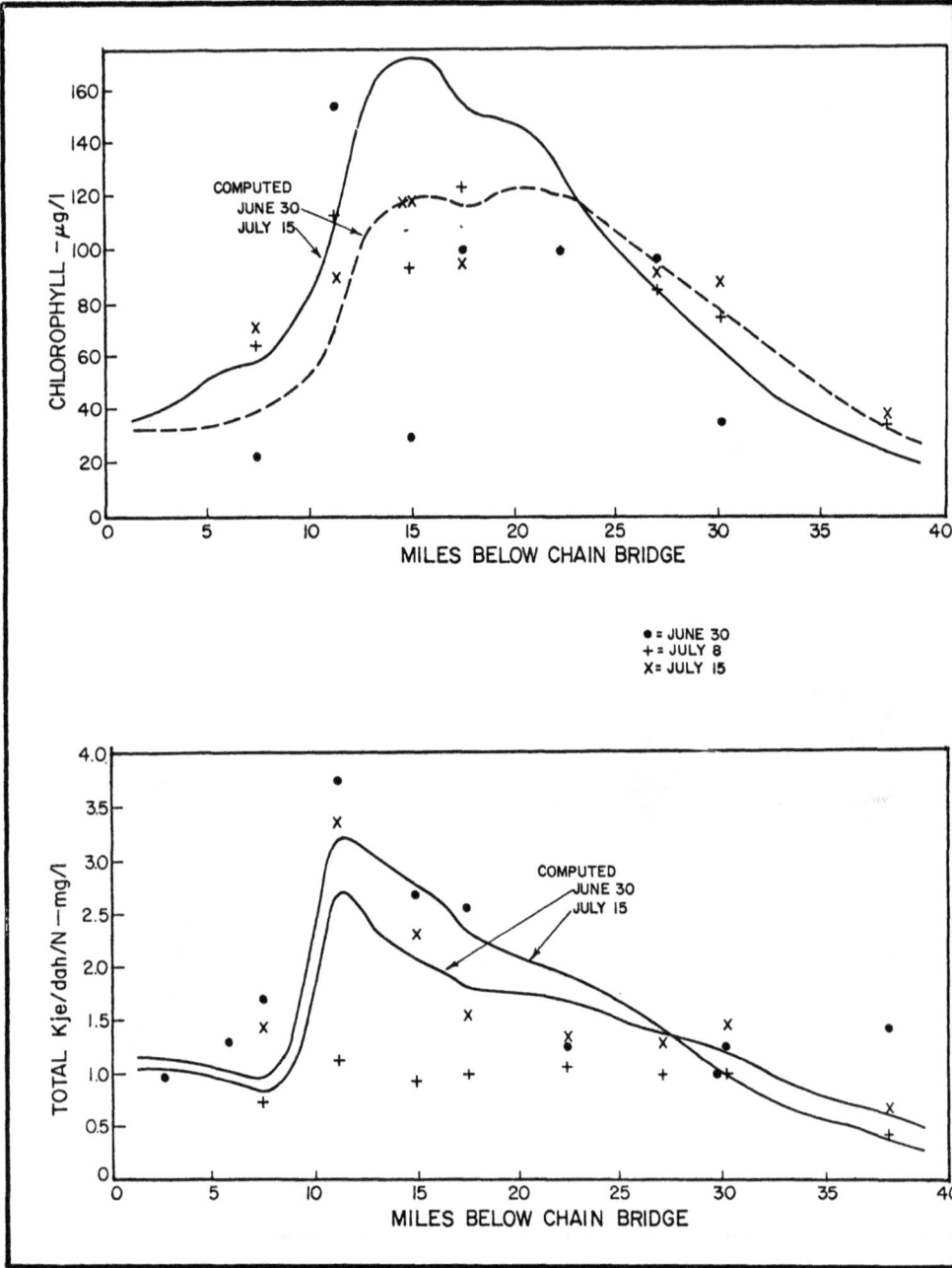

FIG. 11. - SPATIAL VERIFICATION - 1969

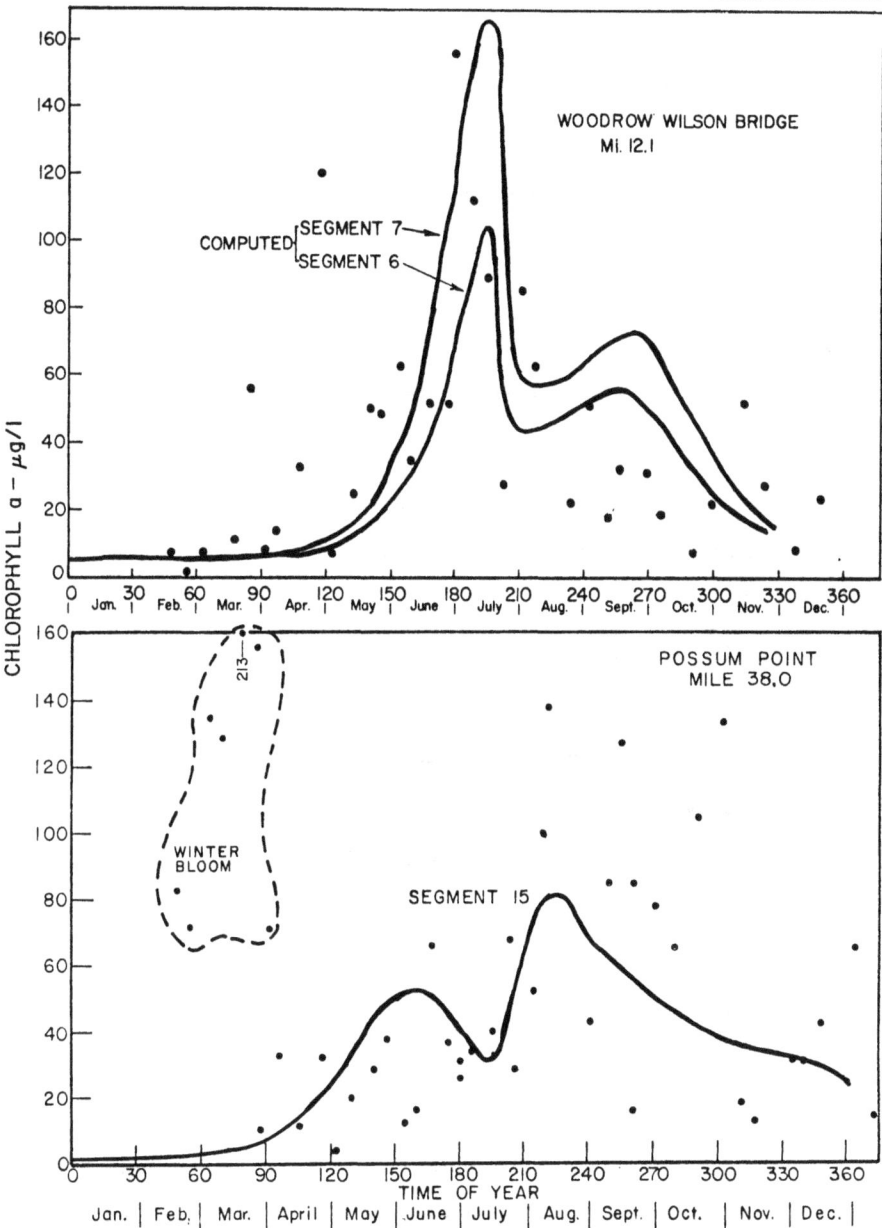

FIG. 12. - TEMPORAL VERIFICATION - 1969

with the data to approximately 40 miles downstream from Chain Bridge.
Temporal variability in phytoplankton is reasonably reproduced
through the first 20 miles and approximately so in the remaining
20 miles. Transient blooms in the late winter and early fall of
1969 are not duplicated as shown. In spite of these inadequacies,
the analysis does reproduce the peak concentrations in the critical
area immediately downstream from Washington, D.C.

Figure 13 shows the calculated spatial chlorophyll profiles
for June 30-July 15, 1969, and those projected with 90% reduction
of nutrients. For this treatment level, maximum concentration of
phytoplankton chlorophyll in the estuary exceeds 100 μg/l or four
times the objective of 25 μg/l. However, the effect of the removal
program is significant from Mile 25 to Mile 40. In that region,
phosphorus appears to limit phytoplankton growth. No distributed sources of nutrient were included, which would increase the
nutrients available for growth.

Upstream boundary concentrations have an important effect,
especially after waste inputs are reduced by treatment. The effect
of incoming phytoplankton chlorophyll was examined under median
flows for two cases: (1) a concentration of 1 μg/l entering the
estuary from upriver; and (2) the objective concentration of 25
μg/l.

Figure 14 shows the temporal variation in chlorophyll for the
main channel, Segment 9, and its associated tidal embayment, Piscataway.
For the case of 1 μg/l chlorophyll entering the estuary, the effect
of the reduced waste discharge on the main channel is minor, as

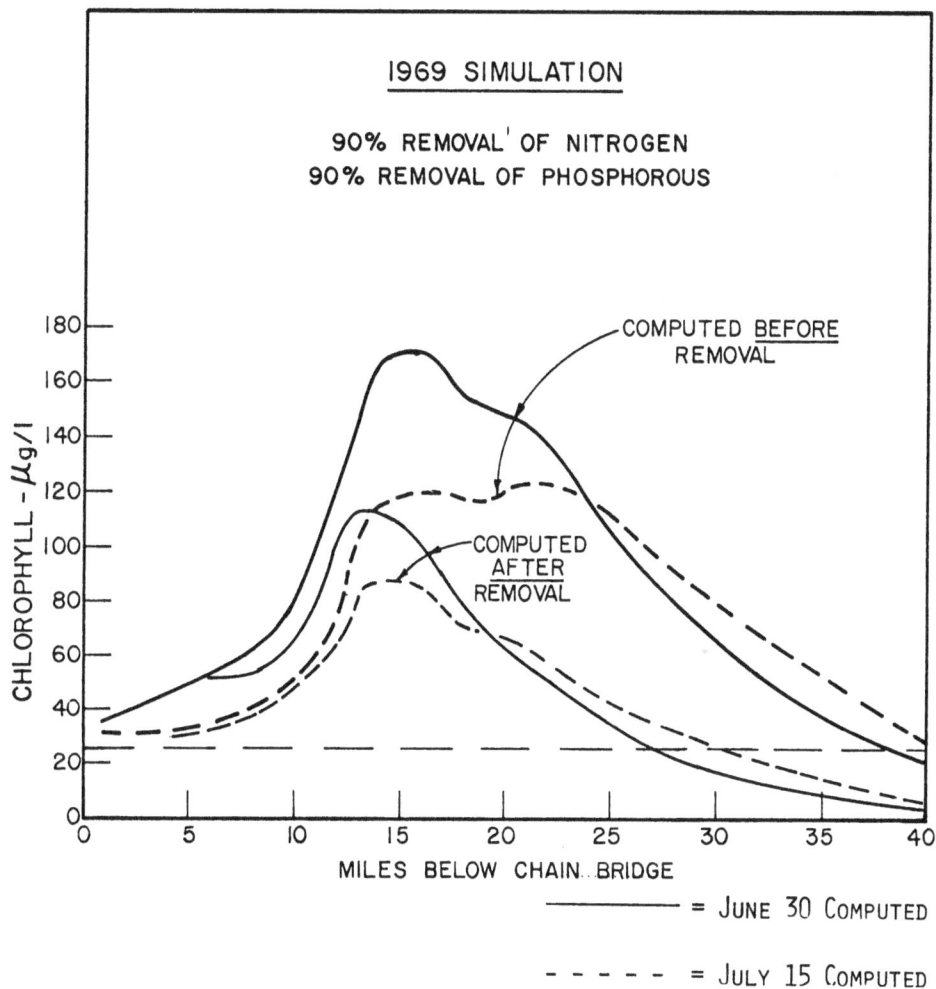

FIG. 13. 1969 SIMULATION OF CHLOROPHYLL

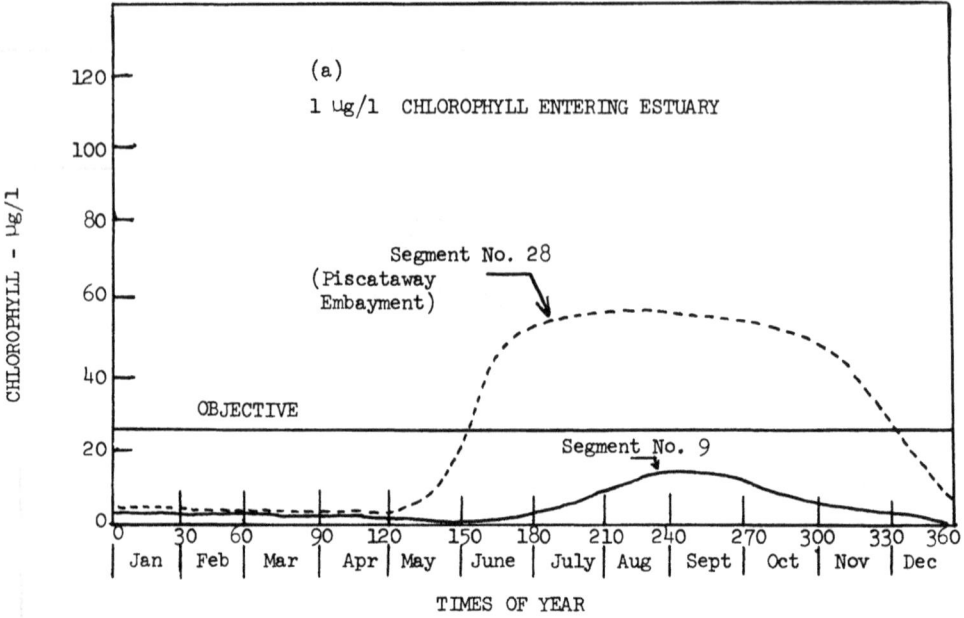

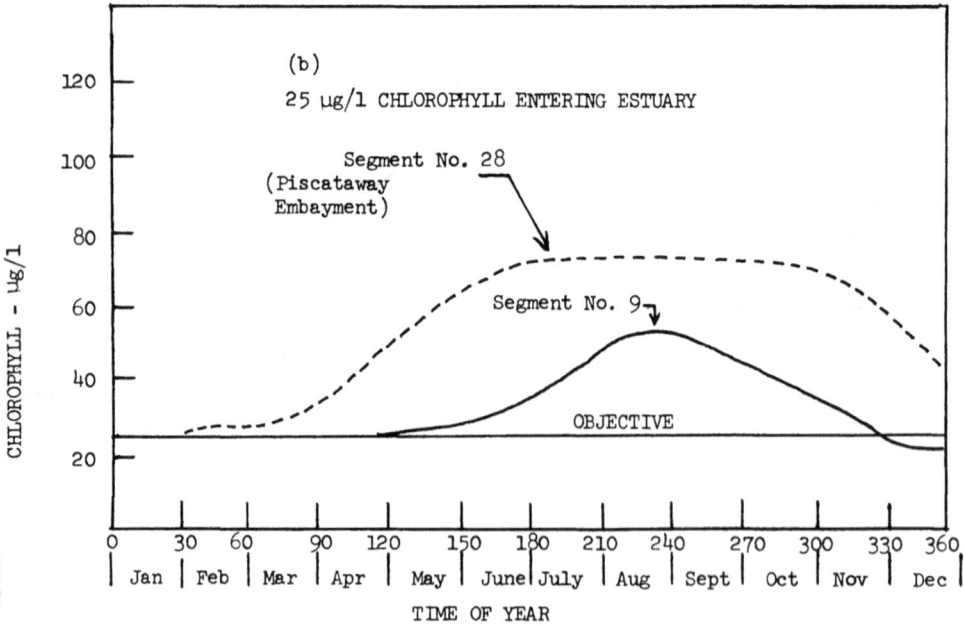

Fig. 14   Temporal Variation in Chlorophyll a at Segments #9 and
#28, Median Flow Simulation. a) 1 μg/l Chlorophyll
Boundary b) 25 μg/l Chlorophyll Boundary

shown in Figure 14a, and results in a summer increase to approximately
14 µg/l chlorophyll. In the embayment section, however, chlorophyll concentrations rise to approximately 55 µg/l for a period of approximately
120 days. The level of 50 µg/l, however, is considerably less than the
concentrations of 150 µg/l - 200 µg/l before nutrient reduction.

Figure 14b shows the effect of boundary chlorophyll concentration at a more realistic level, 25 µg/l. Main channel chlorophyll
levels rise to over 50 µg/l in August and, in the embayment, concentrations rise to over 70 µg/l for a period of about 4 months.

Figure 15 is a longitudinal profile for July 15 for the median
flow and a 25 µg/l chlorophyll boundary condition. For an approximately 25-mile region of the estuary, the concentration of phytoplankton chlorophyll exceeds the objective and rises to a maximum
value of almost 60 µg/l. It is nevertheless a reduction of approximately 60% from maximum chlorophyll concentrations before nutrient
removals.

These projections permit the following general observations.
Achievement of the objective of 25 µg/l chlorophyll in the estuary
may not be possible, primarily because of the effect of discharges
from the Potomac River into the estuary. Even under a median flow
regime (4,000 cfs) during the growing season, maximum concentrations
of 50 µg/l chlorophyll are calculated for the main channel and 75
µg/l in some embayments. On the basis of this preliminary model,
nutrient reduction of 90% may provide reductions in average chlorophyll

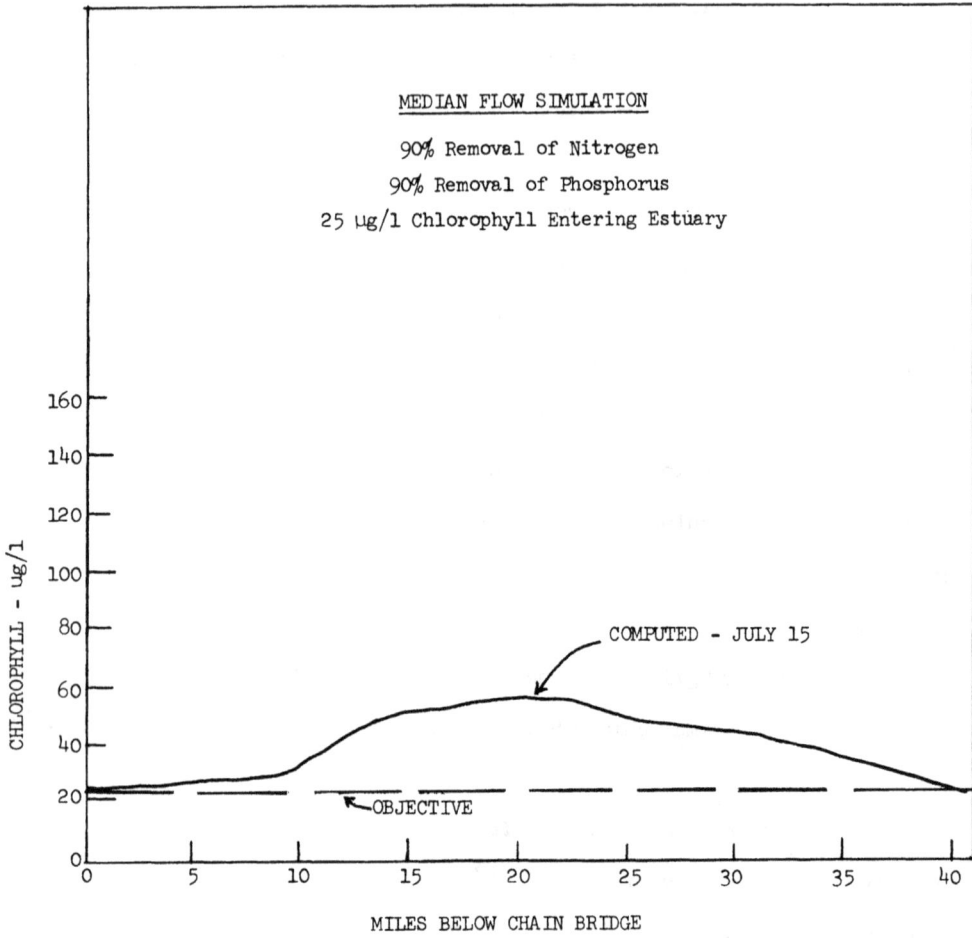

Fig. 15 Median Flow Simulation Profile of Chlorophyll

levels of approximately 60% under median flow conditions in the Potomac.

WESTERN LAKE ERIE

The purpose of this section is to indicate the approach taken on the analysis of the eutrophication problem in lakes, in which water quality differences exist between the littoral and central zones of the water body. Emphasis is therefore placed on the horizontal distribution of the constitutents, rather than on the vertical variation. The approach taken is particularly suitable to relatively shallow lakes in which the vertical differences may be appropriately averaged. The western basin of Lake Erie is generally characterized by such features.

The basin receives the flow from its tributary drainage areas, primarily from the Detroit and Maumee Rivers, and the wastewater effluents from a number of municipalities and industries. The transport within the system is affected by these river flows, dispersion within the basin, and exchange with the central portion of Lake Erie. Figure 16 shows the segmentation of the system (5 littoral and 2 central segments), which permits a realistic portrayal of the transport phenomena, as well as reasonable representation of the concentrations in both the littoral and central zones of the western basin. The advective components of the transport, as shown, are affected primarily by the flow of the Detroit River through the basin. Field observations as well as a hydrodynamic model of the system provided information on the magnitude of this velocity

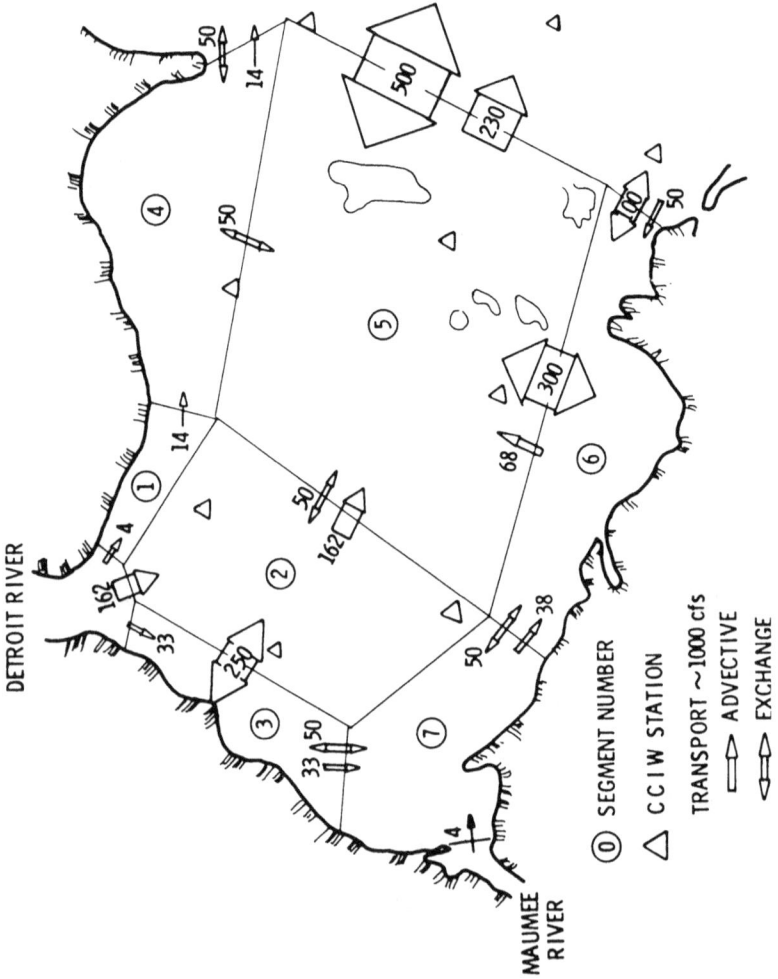

Fig. 16. Map of Western Lake Erie segments, transport components, and stations.

field. The horizontal distribution of a conservative tracer, the chloride ion, provided the basis for assignment of appropriate values to the dispersive components. The components of the transport mechanism have been converted to a volumetric basis, as shown in Figure 16, and are assumed to be at steady-state over the late spring to fall period.

The major components of the kinetic system, as shown in Figure 2, are incorporated in the analysis with the exception of the herbivorous zooplankton. Three nutrients (ammonia, nitrate, and phosphorus) are included in the growth equation, with a preferential assimilation by the phytoplankton of ammonia over nitrate. The form of nutrient equations is a simple product of the Michaelis-Menton type. Two sources of detrital organic nitrogen are considered: (1) the organic nitrogen produced by phytoplankton and zooplankton endogenous respiration and; (2) the organic nitrogen equivalent of the grazed but not metabolized phytoplankton excreted by the zooplankton. The nitrogen that results from these processes is not completely recycled into the non-living organic nitrogen system because the data indicate a substantial loss of total nitrogen from the Western Basin. It is hypothesized that this loss is due to settling of the particulate fraction of the total nitrogen.

In accordance with the general procedure outlined in the introduction, the calibration of the model was effected using 1968-1970 data. The boundary conditions and inputs are associated with the Detroit and Maumee Rivers and tributary streams, and with water

quality conditions in the main body of the lake. The parameters
which resulted from the calibration procedure are presented in Table
1. Comparison of the computed output and observations for both
nutrients and chlorophyll is shown in Figure 17. The temperature
and radiation distributions are also shown.

The verification of the model was effected by an analysis of
water quality data collected during 1930. The parameters were adjusted
to reflect temperature, light and extinction conditions at that time.
The nutrient inputs were also reduced to reflect the population and
development existing in that year, particularly with respect to
phosphorus, since detergents were not in use. The fit between the
computed 1930 distributions and the observations at that time were
comparable to those of the 1968-1970 period.

The model was used to assess the future effect of various levels
of nutrient control. The nutrient inputs to the Western Basin were
projected based on an increase of urban runoff and municipal and
industrial contributions in accordance with the population increase
shown in Figure 18. This figure also presents the data on phytoplankton counts since 1920 for comparative purposes. Natural runoff
and agricultural drainage were assumed to be constant for the period
of the projection. All other external variables were held at 1970
values. Under a policy of no control of nutrients, the projection
is presented in Figure 18. The chlorophyll concentration shown
is the summer average for segment 7, adjacent to the Maumee River.
A currently favored control policy is aimed at removal of the phosphorus. Projected conditions for both an 80 per cent removal policy

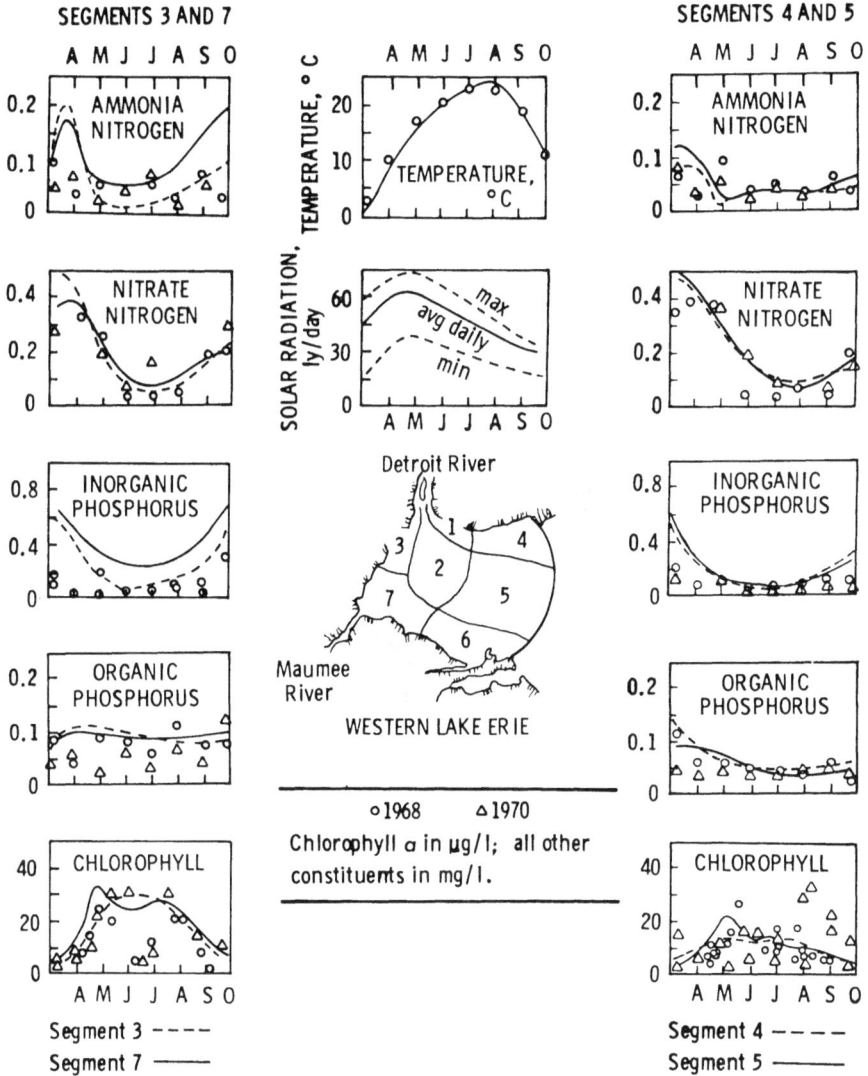

Fig. 17. Calculated profiles versus observed data.

LAKE ERIE WESTERN BASIN

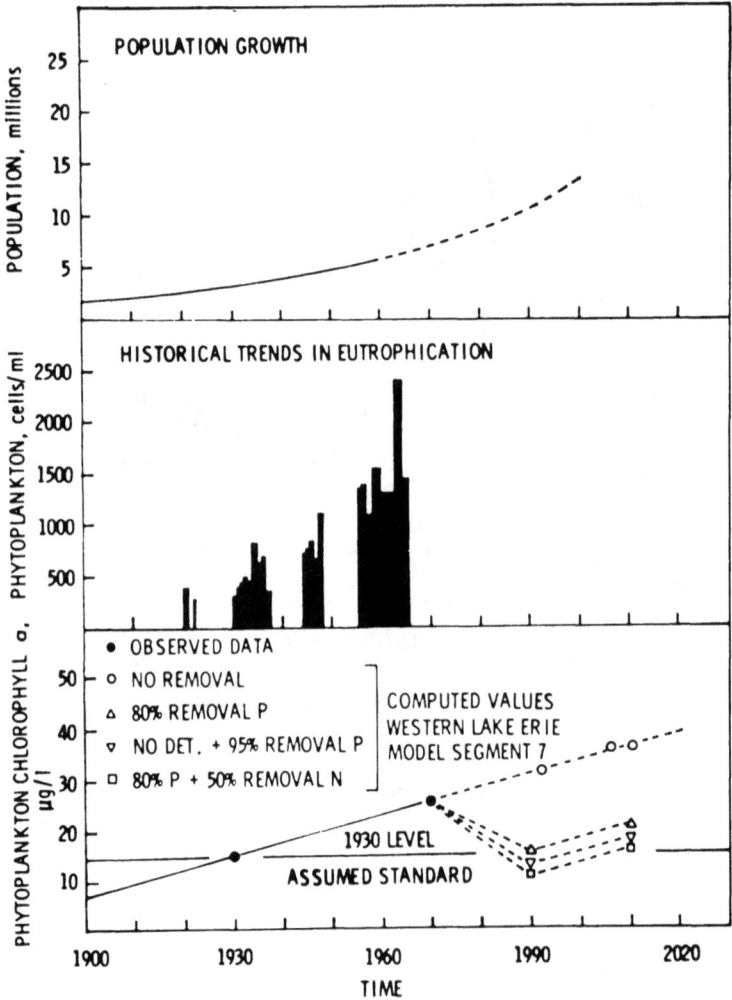

Fig. 18. Projections.

in addition to a total ban on detergent phosphorus are also shown
in Figure 18. It is important to note that the level of removal
assumed for the more stringent control policy may not be technologically
feasible. An alternative policy that appears to be feasible using
presently available technology is to remove 80 per cent of the phosphorus and 50 percent of the nitrogen being directly discharged to
the basin. The projected result is shown in Figure 18.

Other planning alternatives were investigated: the effect of
changes in lake level on eutrophication (a shallower body of water is
generally more productive than a deeper one) and an agricultural land
use policy which results in a 50 per cent decrease in the phosphrous
content of the agricultural runoff. Each of these two planning
alternatives increased the projected phytoplankton level by no more
than 10 per cent. This variation is within the probable error of
the projections so that the precise magnitude of the effect is in
doubt, although it is likely to be small.

LAKE ONTARIO

The analysis of phytoplankton growth in lake systems which are
characterized by significant vertical stratification is the main
purpose of this section. By contrast to the approach taken in the
previous example, emphasis is placed on the central, rather than
the littoral, zone of the lake. The central section of Lake Ontario,
a large lake, is used as an example. This analysis provides a preliminary basis for estimating the direction of change for various
control programs.

There are particular problems associated with the water quality of a large lake. Because of the size of such systems, immediate improvement is not observable after treatment facilities are installed or control action is taken. The difficulty of obtaining representative samples of these systems is evident. This condition applies to the spatial and temporal distribution of water quality constituents, biological components, and hydrodynamic circulation.

Lake Ontario receives the outflow of Lake Erie through the Niagra River and flow from a number of rivers in its tributary drainage basin. The flow from Lake Erie is 90 per cent of a total annual outflow of approximately 230,000 cfs. By contrast the mass inputs of nutrients are about equally divided between the Lake Erie inflow and the direct inputs to Lake Ontario from municipal and industrial sources and tributary rivers.

The basic physical features included in the analysis are shown in Figure 1D. The model is divided vertically into the epilimnion, hypolimnion, and benthos, each of which is assumed to be of uniform concentration. Mixing in the vertical direction is effected during isothermal conditions and is restricted during the summer to simulate vertical stratification. The kinetic interactions incorporate all of the components shown in Figure 2, including settling. The discharge data were obtained from the International Joint Commission, and are categorized as municipal, industrial, and tributary. These loadings are used as boundary conditions and forcing functions for the nutrient system.

Comparisons of the computed profiles and the observed data
are shown in Figure 19, for the phytoplankton, zooplankton and
nutrient systems. The spring growth phase to approximately mid-
May is due primarily to increasing light and temperature. Growth
continues until phosphorus limitation becomes significant in early
June, with minimum predation by zooplankton. During mid-summer,
the nutrient limitation effect and increased grazing account for
the minimum values of biomass in July. At the same time, biological
recycling of nutrients, associated with increased zooplankton grazing
becomes more significant. Nutrients are returned to the epilimnion
through excretion as well as by phytoplankton endogenous respiration.
In late July, due to nutrient regeneration, a second phase of
growth is stimulated. Since nutrients are at low levels, the growth
rate is relatively low through late summer. The fall overturn of
the lake which produces mixing with the colder hypolimnion waters
and the decreasing light further reduces the growth of the phytoplankton.

Figure 20 shows the estimated input of phosphorus to the epilimnion due to recycling by both biological effects and to transfer
by vertical mixing from the hypolimnion. During early spring the
hypolimnion contributes almost twice as much nutrients to the epilimnion as do external sources; and in June-July biological recycling
reaches almost five times the mass rate of external phosphorus sources
due to waste discharges, Niagra River and tributary streams. Zooplankton excretion recycles almost twice the input rate, and in

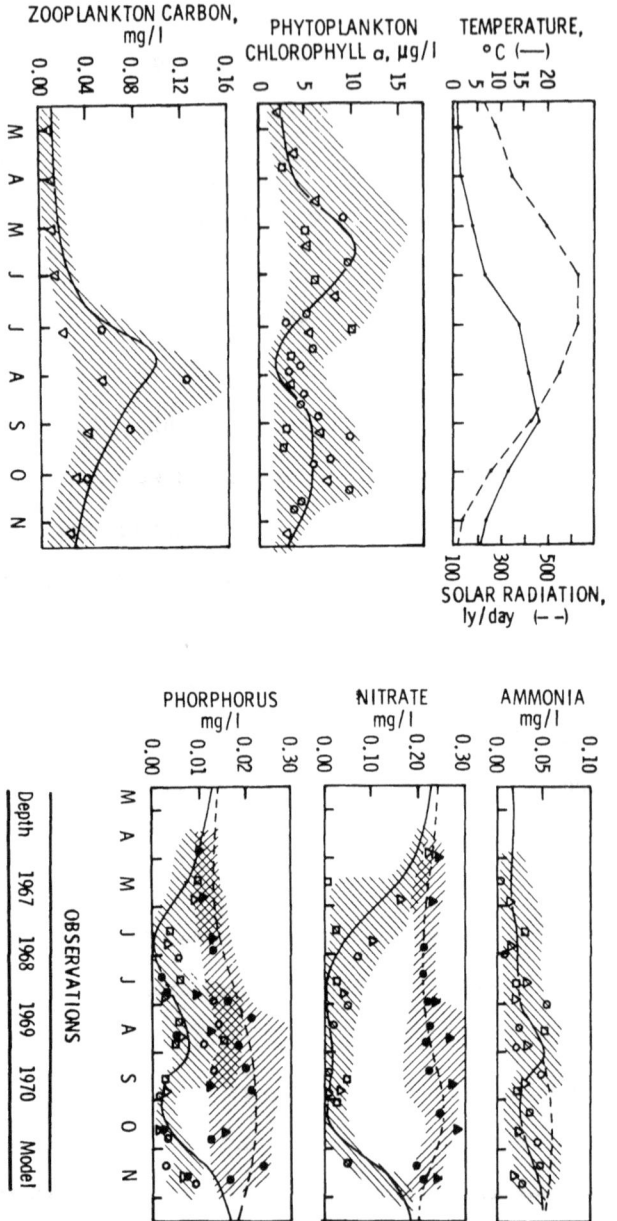

Fig. 19. Comparisons of observations and computed profiles.

Biological Recycling

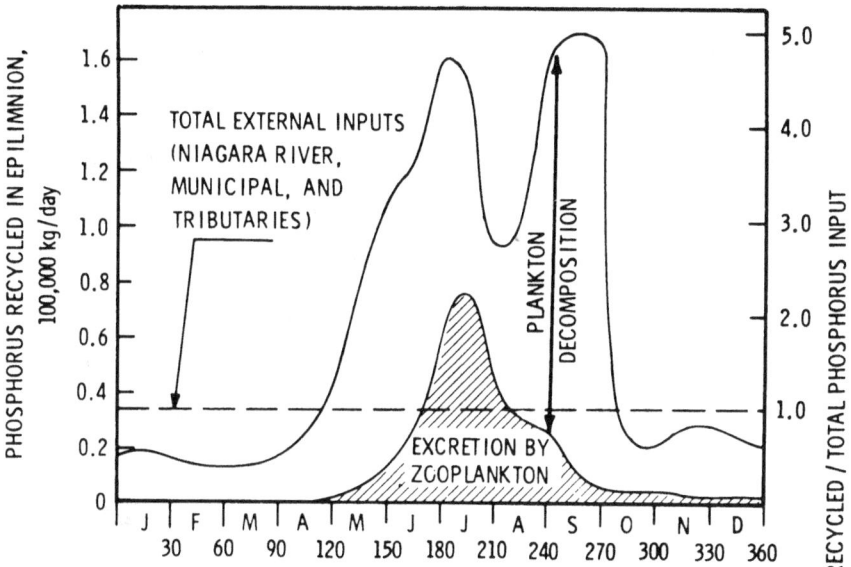

PHOSPHORUS RECYCLED IN EPILIMNION, 100,000 kg/day

RATIO: PHOSPHORUS RECYCLED / TOTAL PHOSPHORUS INPUT

TOTAL EXTERNAL INPUTS
(NIAGARA RIVER,
MUNICIPAL, AND
TRIBUTARIES)

PLANKTON DECOMPOSITION

EXCRETION BY ZOOPLANKTON

Hypolimnetic Recycling

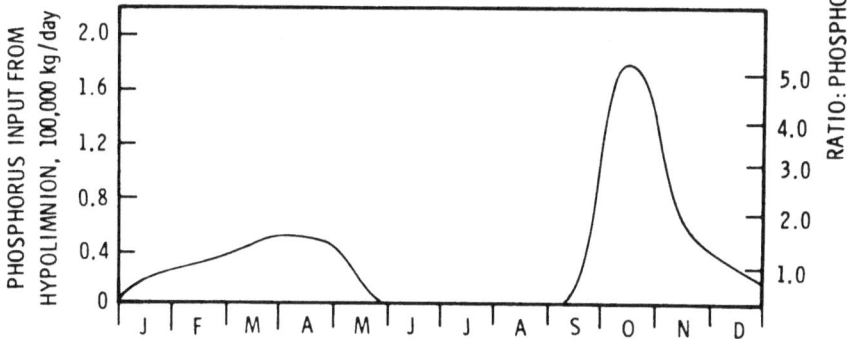

PHOSPHORUS INPUT FROM HYPOLIMNION, 100,000 kg/day

RATIO: PHOSPHORUS RECYCLED / TOTAL PHOSPHORUS INPUT

Fig. 20. - Recycle Effect

late summer -- early fall period, biological recycling reaches
another peak, followed by a rapid decrease due to fall overturn.
The flux of nutrients from the hypolimnion during October is not
sufficient to offset the drop in water temperature in the epilimnion.
Biological recycling, therefore, drops sharply in October and the
phytoplankton biomass declines. The results shown indicate quantitatively the significance of biological recycling and vertical
mixing relative to the level of external nutrients inputs. A
reduction in input phosphorus load, therefore, may not necessarily
result in a concomitant reduction in biomass in the "short run" of
several years. Indeed, additional analyses indicate that 5-10
years may be required for phytoplankton biomass to reach a new
level.

One of the most significant questions which must be answered
for systems such as the Great Lakes is the effect of various measures to control the potential eutrophication problem. Considering
the possible importance of the nutrient recycling, on the one hand,
and the economic cost of control facilities, on the other, this
question takes on added significance. In order to provide some
insight into this problem, consider the following. Figure 21 shows
a time plot of 8 years of solution, equivalent to one detention
time. The conditions are comparable to previous runs but use a
Michaelis phosphorus constant of 10 μg/l instead of 2 μg/l. As
shown, peak phytoplankton levels in the spring gradually decrease
from greater than 9 μg/l in the first year to about 7 μg/l by the
eighth year after which time a dynamic equilibrium is reached.

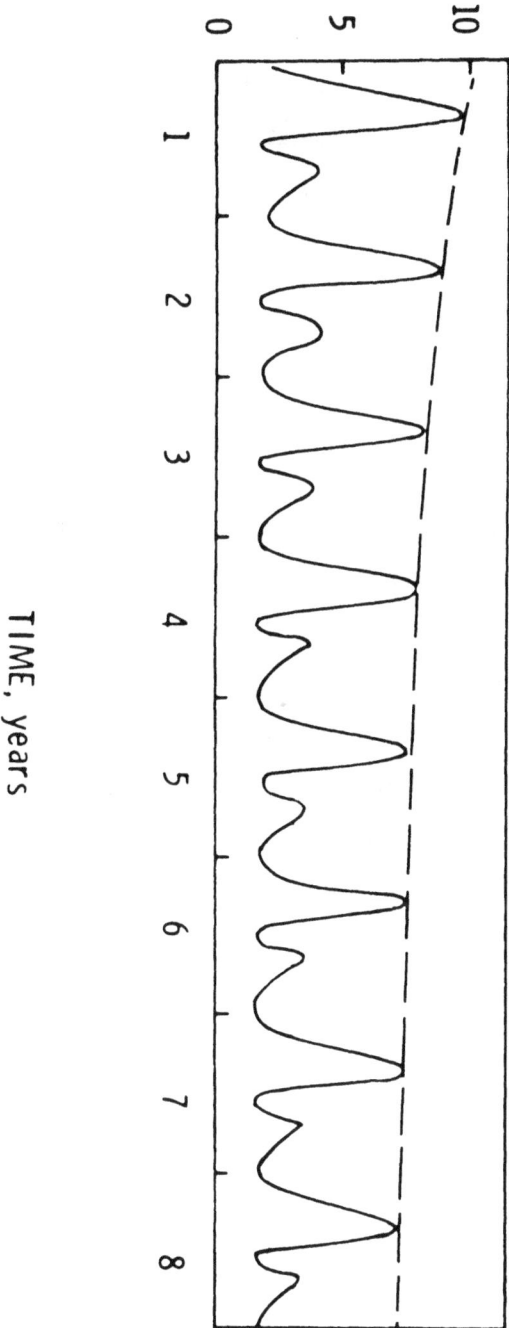

PHYTOPLANKTON CHLOROPHYLL *a*, μg/l

TIME, years

In summary, the analysis bears out two significant points: first, due to the relative magnitude of the internal recycling of nutrients and the impact of the Lake Erie input, reduction of in-basin nutrient discharges does not necessarily produce a one-to-one correspondence in the reduction of phytoplankton levels; and second, in order to determine the effect of nutrient reductions, model runs will have to be extended for a period of about 10-20 years to ensure that a new level of dynamic equilibrium has been achieved.

VALIDATION

The nature of most environmental problems is characterized by time-variable, three-dimensional, non-linear elements. Assuming our scientific knowledge of the real situation in these systems is complete, at least in principle, an abstraction is drawn of this understanding in which the significant factors are included and the irrelevant features are omitted. If not omitted, they are frequently modified. This process of abstraction in constructing the model therefore necessitates its validation to insure that the relevant factors have been realistically included.

The process of validation for water quality models involves comparisons between model output and past observations. When this comparison is satisfactory, the model is said to be validated. What constitutes "satisfactory" depends on the nature of the problem, the structure of the model, and the extent of available data.

The highest level of validation, which may even justify use of the term "verification," is achieved when a model prediction is reproduced by observations in time and space. Such predictions are the goals of many applied and fundamental sciences, such as astronomy and meteorology. In such cases, one has an absolute basis of validation -- direct comparison of an observation with the prediction of the model. This type of validation in environmental modeling is, in any practical case, almost impossible to achieve because of the time lag involved and equally important because of the probablistic character of the hydrological, meterological, and waste inputs.

Water quality models, on the whole, are thus planning tools, rather than predictive techniques. The model is used primarily to examine the spectrum of responses of the system which may occur under varying planning alternates. In many cases, therefore, extreme accuracy is not required, but rather trends or directions may be sufficient to answer the planning problem. It is thus the nature of the problem which primarily determines the degree of validation required. The degree of certainty required is determined by many factors the most important of which is the consequence of an erroneous decision.

The nature of the problem and, specifically, the time and space scales of the problem, dictate the degree of simplicity or com-plexity of the analysis. Given or assuming these scales, a specific question is posed and the purpose of the analysis is to answer this question in the simplest, most efficient and most realistic

manner. For example, if an analysis of the long-term build-up
of conservative substances in a lake is required, a simple completely mixed model with a time interval of one year may be ten-
tatively assumed as a basis for an adequate analysis. By contrast, if the question concerns the bacterial distribution at a
bathing area due to storm runoff, it is evident that a model of
smaller time scale is required -- a time variable analysis in
two or three dimensions.

Furthermore, many important water quality problems in natural
systems can and have been answered by a steady-state analysis with
linear kinetics and simple transport components. The simple
dissolved oxygen models of streams and estuaries are typical examples.
Such problems may be approximated with sufficient accuracy to
yield an analysis which is adequate for decision-making regarding
the level of treatment of wastewater. On the other hand, at this
stage of development, an analysis of the eutrophication problem
usually requires time variable and non-linear terms in order to
determine the effect of nutrient removal from wastewaters.

The relative complexity of the model is an important factor in
its validation -- the more complex, the greater the degree of validation required. The complexity of the model is measured by the
number of the transport, kinetic and input terms in the equations.
Compare the simplicity of the equation used to describe the
steady-state distribution of a conservative substance in streams to
the complexity of those describing the time-variable distribution

of nutrients and phytoplankton in lakes. For each additional component which is included in the analysis an additional degree of validation is desirable, indeed necessary. As a consequence, additional data are required. If data are not available on a specific component, it is questionable whether it should be included in the analysis. If it neither increases understanding nor expands the area of application, is is a futile exercise. It may admittedly develop computational skills and facilities, but this is usually not a sufficient justification for increasing the complexity of a model in answering a problem.

Having selected what one believes to be an appropriate model based on the time and space scales of the problem, the next step is the analysis of data which are representative of these scales. This step frequently involves a spatial and/or temporal averaging of the available data to determine the mean and some measure of its variation. Such variations invariably exist due to the stochastic nature of environmental phenomena as well as to the extent over which the integration is taking place. A comparison of these variations with those which the analysis is attempting to address is highly significant.

Following the example of the long-term build-up of a conservative substance given above, the variation of concentration during a year over the lake is compared to the increase in the mean concentration which has been observed or may be expected to occur in one or two decades. Assuming the latter exceeds the former, the

selection of the type of model is more firmly, although not completely, established. If the former exceeds the latter it is
indicative of an unrealistic model represention, of poor selection
of time and space scales, or of inadequate data. If this is the
case, a restructuring of the model is usually called for involving
a greater degree of complexity, i.e. a more refined time and/or
space scale. Assuming it is possible to make the appropriate adjustments, one can then proceed to the actual validation procedure.

The first step in the procedure is the calibration of the model.
Given the external parameters of the system, an estimate is made
of the appropriate transport and reaction coefficients. These may
be determined from a fundamental analysis relating to the specific
coefficient, as may be accomplished in the case of the hydrologic or
hydrodynamic terms, or from a statistical correlation of the coefficient, as is usually done with the biological and chemical kinetic
terms. In any case, assuming a range of these values is known,
a best estimate is made of each, the model is run and the output
compared to the data. Invariably, successive adjustments are required to obtain a "reasonable" fit of the model and data. This
procedure is the calibration stage.

Having established a set of coefficients representative of
one set of external conditions (e.g., with respect to temperature,
flow and waste input) the model is rerun for a different set of
input conditions. If the output agrees, in a "reasonable" qualitative way, with the second set of data, the model is considered

to be validated in the first degree. In some instances, if sufficient data are available, a quantitative comparison is possible. Then, a measure such as the standard error of the mean is used and the model is considered validated if the model output falls within ± 1 standard error. Additional comparisons with different combinations of the exogenous parameters yields higher degrees of validation.

The extent to which this procedure is carried on rests, in large measure, on the conditions of the projections and the environmental consequences of error. Thus, projections to be made representing flow and temperature conditions within the observed range require a lesser degree of validation than those which are based on extrapolations far beyond the observed range. With respect to the second criterion, it is apparent that an analysis which relates primarily to the health and well-being of people necessitates a greater degree of validation than does one which is mainly concerned with aesthetic features of the environment. Thus the degree of validation, taking into account the complexity of the model and the available data, should be evaluated in balance with the consequences of not meeting the environmental criteria for the various planning alternates. Such an evaluation permits the appropriate factor of safety to be incorporated in the planning. Finally, cognizant that such projections are but one of a number of inputs which the decision maker weighs, the analyst must attempt to balance the time, effort and money spent on validation with the other bases on which decisions are made.

Questions invariably are raised about a procedure such as outlined above and particularly about the validity of the model. The question of the validity of the model may be addressed on two levels. First, and most direct, an analysis of the response is conducted after the control is exercised, e.g., after the construction of a waste treatment plant. This procedure should be followed more than it is at present. In cases such as large lake or basin systems which require long response periods, 10 to 20 years, it is impractical. The second level is essentially the one outlined above, based on data previously collected rather than on data to be collected. Ultimately, the validity of such an environmental model is determined by the judgment of the scientific-engineering community on the available models and practicality of their use in addressing a specific environmental question.

* * * * * * * * * * * *

ACKNOWLEDGEMENT

The assistance and cooperation of Walter Matystik of the Manhattan College Research Staff is gratefully acknowledged.

REFERENCES

1. Di Toro, D.M., O'Connor, D.J., and Thomann, R.V., A Dynamic Model of the Phytoplankton Population in the Sacramento-San Joaquin Delta, Advances in Chemistry Series, No. 106, American Chemical Society, 1971, pp 131-80.

2. "Mathematical Model of Phytoplankton Population Dynamics in the Sacramento-San Joaquin Delta," Hydroscience Report to Department of Water Resources, State of California, Sacramento, California, January 1972.

3. O'Connor, D.J., Di Toro, D.M., Mancini, J.L., Mathematical Modelling of Phytoplankton Population Dynamics in the Sacramento-San Joaquin Bay Delta. Presented at Modeling of Marine Systems NATO, Ofir, Portugal, June 1973.

4. Thomann, R.V., Di Toro, D.M., O'Connor, D. J.,"Preliminary Model of Potomac Estuary Phytoplankton," Journal EED, ASCE, June 1974.

5. "Limmological Systems Analysis," Hydroscience Report to Great Lakes Basin Commission, Ann Arbor, Michigan, 1973.

6. Di Toro, D.M., O'Connor, D.J., Mancini, J.L., Thomann, R.V., "Preliminary Phytoplankton-Zooplankton-Nutrient Model of Western Lake Erie." To appear in Systems Analysis & Simulation in Ecology, Volume 3, Academic Press.

7. Thomann, R.V., Di Toro,D.M., Winfield, R.D., O'Connor, D.J., Mathematical Modeling of Phytoplankton in Lake Ontario, Report to National Environmental Research Center, EPA, Grosse Ile, Michigan, 1974.

Fish Population Models: Potential and Actual Links
to Ecological Models

W. E. Schaaf

Fish population dynamics models, characterized as predator-prey volterra systems, are considered a subset of more general compartmental ecosystem models. Both types of models are homogeneous, in the sense that their parameters do not reflect environmental variability; they are equilibrium centered and, therefore, have limited usefulness for predicting the effects of perturbations. Fishery models are shown to aid ecological modeling by 1) scaling the system (i.e. providing a minimum rate of output from a food web), and by 2) helping to specify the functional nature of feeding links, which may be the source of ecosystem stability or instability. Ecological theory and the development of general models will have an impact on fishery population dynamics models through a realization of a more holistic view, that the dynamic behavior of components (e.g. a fishery) is determined largely by the structure of the ecosystem and may be relatively insensitive to the parameters used in the component model. Both fishery biologists and systems ecologists need to develop a broader view that considers alternate steady states.

Fish Population Models: Potential and Actual

Links to Ecological Models

Introduction

The title of this paper implies a dichotomy which I believe is false, but it does offer a convenient point of departure for expressing some fairly general, review comments about modeling biological systems. I hope to demonstrate that there is no real difference in kind between fish population models and ecological models, but only one of degree. An intimate feedback relationship exists between the two that can have a positive effect on the acquisition of knowledge that we resource managers can use to manipulate ecosystems for the long-range benefit of humankind. I will focus on the similarities, the symbiosis of the two approaches, and the potential links requiring further research.

In order to do this I'll first categorize population dynamics models, which is relatively easy to do, and then try to categorize ecological models, which is more difficult because they are so diverse. This discussion will be restricted to one particular class of ecological models, which will never-the-less demonstrate that fishery models are a subset - and, in my opinion, a very important subset - of ecological models.

Most model taxonomies consider details of the mathematical structure or assumptions employed; for example, discrete vs. continuous, deterministic vs. stochastic, linear vs. nonlinear. For our discussion I think it will be useful to consider more abstract, qualitative features of models, such as where they fit in Levins' (1968) scheme of maximizing

either realism, generality, or precision, or whether they are homogeneous or not. Here homogeneous is not mean in the mathematical sense
(i.e., $F(tx, ty) = t^n F(x,y)$), but in the structural sense of whether
or not a neighborhood stability analysis correctly describes the global
stability of the system (cf. May, 1973, p. 15). That is, do we have an
equilibrium centered view, or are we concerned with the transient behavior of systems (Crutchfield, 1973) that are not near the equilibrium
state and for which the probability of structural changes (i.e., extinction of species) may not be small (Holling, 1973)?

Fish population dynamics models

Fishery research encompasses a broader area than the type of
modeling discussed here. In a general way such research can be divided
into two branches (Cushing, 1968). One branch deals with the natural
history of fish, their growth, reproduction, genetics, physiology and
behavior - as it relates to migration, inter- and intra-specific competition, etc. Each of these areas of ichthyology has developed quantitative techniques and syntheses. That is one form of modeling.

The other branch, more germaine to our discussion, is concerned
with the dynamics of exploited populations. It had its foundation in
watching many great stocks of fish be depleted as harvesting became more
intense and efficient. This aspect of fishery research, which has developed a cohesive body of observation and theory, "the modern theory of
fishing", has resulted in a rather narrow class of mathematical models
intended to describe and predict the response of a single species to
various rates of exploitation.

Two basic approaches have evolved to modeling the dynamic response of a fish population to harvesting. They are not conceptual alternatives, but rather different ways of solving the same general model; they result from different types of data and simplifying assumptions. The general model may be constructed by expressing the relative rate of change in biomass of the exploited population as the sum of input rates and loss rates to the biomass (Schaefer and Beverton, 1963):

$$\frac{dB}{Bdt} = r(B) + g(B) - M(B) - F(E) + e \tag{1}$$

where r, g, and M = rates of recruitment, growth, and natural mortality, and are functions of the biomass (B) present. F = fishing mortality rate, and is a function of the amount of fishing effort (E). e is a random environmental variable.

The first approach, and most widely used one, has been to estimate separately each of the basic rates in equation (1) and, by making certain assumptions, combine them into special forms of the general model. By focusing attention on the steady state, in which the biomass is not changing, the catch can be expressed in terms of r, g, M and especially F which is somewhat under the direct control of man through regulation of his effort. This "analytical" approach was the first (Baranov, 1918) and most widely used (Thompson and Bell, 1934, Ricker, 1944); it has been most completely developed by Beverton and Holt (1957) and is frequently referred to as the Beverton-Holt model.

The other approach, which was most completely developed, and applied to fishery problems by Schaefer (1954), combines the elemental rates of recruitment, growth, and natural mortality, and expresses the rate of population increase as a single function of population size,

f(B). The catch rate at equilibrium can then be expressed, from (1),

as:

$$F(E) = f(B) \qquad\qquad (2)$$

Various assumptions can be made about the form of f(B), one
of the simplest being that it is linear with B, so

$$f(B) = K \ (L-B)$$

where L is the maximum population size the environment will support.
This is the well known "Logistic" law of population growth, discovered
by Verhulst in 1844; the resulting fishery model is also sometimes
referred to by this name. The usual assumption for $\bar{F}(E)$ in equation (2)
is that it is proportional to E (i.e., F(E) = QE). These two assumptions
lead to a parabolic relationship between catch and effort.

Both strategies for applying equation (1) have traditionally
ignored e, the environmental variable; though Watt (1968) has discussed
the effect of density independent factors, such as weather, in terms of
how much scatter is produced about the expected equilibrium catch parabola.
Presumably, the effect of a deleterious factor added to the environment
could be viewed in much the same way, that is how it shifts the equilibrium catch curve, but this would not provide much a priori, analytical
power to predict the dynamic response of the population to sudden changes
in the environment. A more useful (though more data hungry) approach
might be to retain the elemental rates, as in the Beverton-Holt model,
and express them as functions of environmental variables, in addition to
just biomass. Recent developments of the Leslie matrix approach (Leslie,

1945) show great promise for relating environmental effects to population dynamics. The elements of the population projection matrix,
age-specific fecundity and mortality rates, can be replaced by functions
of population size (Usher, 1972). Jensen (1971) has applied the matrix
approach to simulate changes in the yield from a brook trout population
that result from a toxicant that increases the mortality rate of juvenile
fish.

Ecological models

Ecological models have been used primarily to describe distribution of plants and animals in space and time. Ignoring discussion of
physiological autecology (as we ignored the natural history aspect of
ichthyology), early descriptive field studies of plant or animal aggre-
gations resulted in attempts to fit statistical frequency distributions
to observed patterns of relative species abundance (Williams, 1964).
Closely associated with the species aggregation problem were the many
attempts to describe mathematically the total diversity in biological
communities, and the spatial distribution of component populations (Hairston, 1959). The recent and elegant theory of island biogeography
(MacArthur and Wilson, 1967) is an extension to larger regions of this
type of description of species diversity patterns. Application of life
table techniques, from human actuarial studies, also occupies a venerable
place in the study of population growth (Deevy, 1947).

The type of model that most of us refer to now-a-days, however,
as an ecological model - certainly as evidenced by the papers at this
symposium - had its inception with the paper of Lindeman (1942), out-

lining the trophic-dynamic aspect of ecosystems. By the mid-'60's many compartmental models were developed (Clark, 1946, Harvey, 1950, Odum, 1957, Teal, 1957, 1962, Mann, 1969), in which ecosystems are viewed as consisting of interconnected compartments (sometimes of several component species) and interest centers on the amount of energy or material in each compartment and in the flow rates between compartments. The ecological processes going on within the compartment among various species are assumed to "average out" into some fairly simple behavior for the whole compartment. By 1966 Holling had formally proposed an approach to modeling ecosystems that focused attention on the functional nature of the rate processes (e.g. by experimentally determining the form of the predator-prey interaction) and "building up" from there. Interest still centers on dynamic behavior of ecosystems. Recent ecosystem models usually use some mixture of the compartmental and the functional components approach (Van Dyne, 1969).

Relationship of fishery and ecological models

Attempts to explain the variation in numbers of a species have led to a class of ecological models, the coupled differential equations of competition and predation of Volterra (1926), or the similar Ross malaria equations (1911) which were some of the earliest in the ecological literature. These are the basis for the Schaefer-type fishery models. It should be apparent from the foregoing discussion that an intimate relationship exists between fishery models and ecological models. Figure 1, from Levins (1968), illustrates this relationship.

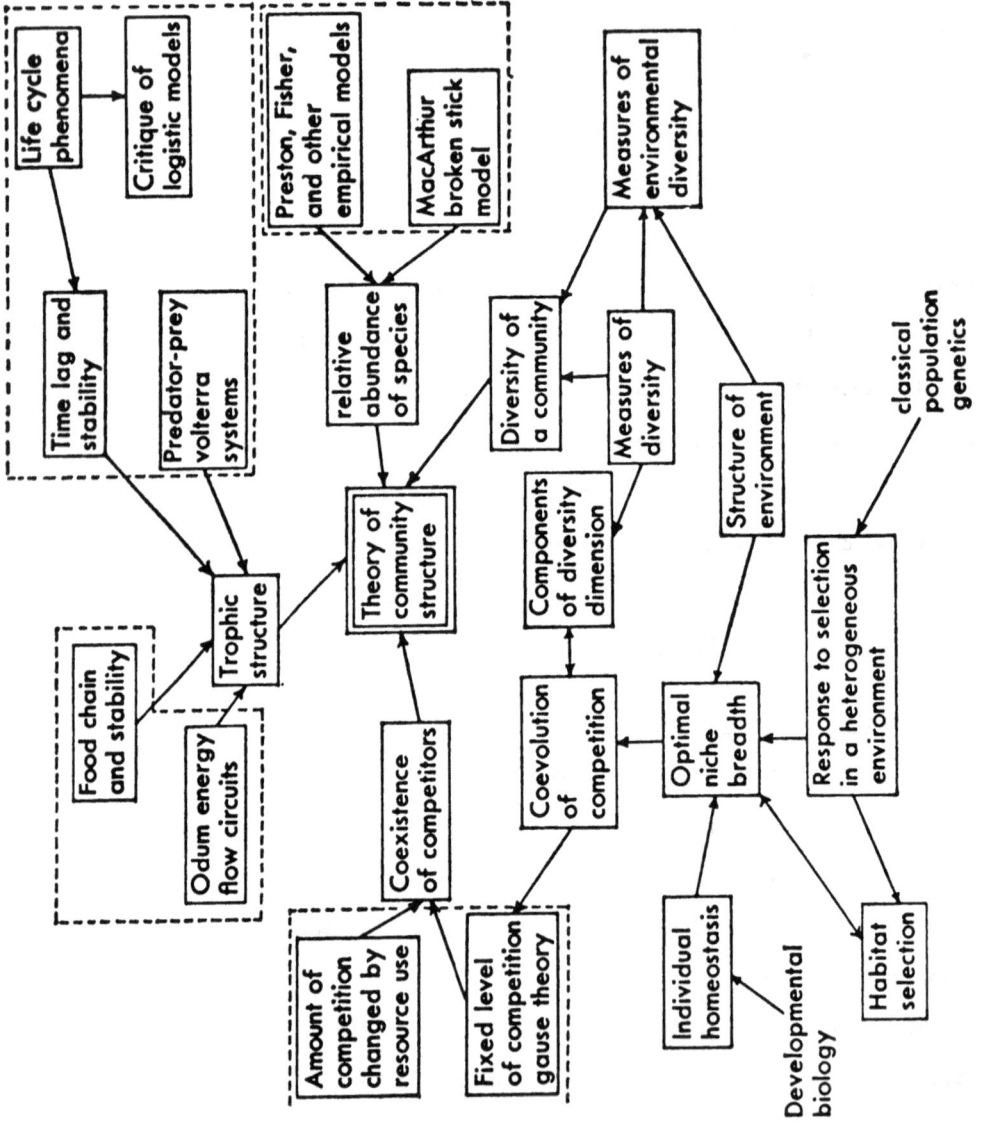

Figure 1. Relations among some of the components in evolutionary population biology theory.

Fishery models, which fall in the upper right hand box, feed directly into the "trophic structure box," which encompasses the ecological models being considered here. The compartmental ecological models are attempts to understand in general terms the structure and functioning of entire ecosystems. They are usually not particularly realistic or precise and are not intended for application to specific environmental problems. Fishery models, which in a sense follow Holling's strategy of modeling the functional process of one component of the ecosystem, usually sacrifice generality for a realistic and precise model applicable for management purposes.

Both ecological models and fish population models, as usually conceived, are homogeneous in the sense (Smith, 1972) that the parameters do not vary in relation to temporal or spatial variations in the environment; the space occupied by the system is considered homogeneous with respect to the processes (biological interactions) going on so that average values for parameters are sufficient to represent any spatial or temporal variability. Another way of making the same characterization of these models is by use of Holling's (1973) concepts of resiliancy vs. stability, and domains of attraction. Our models tend to be equilibrium centered and provide little information about transient behavior of systems not near the equilibrium point. They don't predict, or even consider, alternate stable points (Sutherland, 1974). As such, they tend to work - only so long as they work This is because the models are not necessarily descriptions of the underlying biological processes. They are made to fit natural changes in the recent past and do not consider the effects of altered ecosystems, such as result from pollution. Our model systems are not stressed, and we do not have a very clear idea about where the boundaries are. Examples will be given later.

Fishery input to structure and function of ecological models

Fishery biology has traditionally been more ecosystem oriented than most other areas of resource management, perhaps because of the emphasis in commercial fisheries on biomass, and hence on the production process (Wagner, 1969). Ecological theory has always played an important role in the development of fishery biology, but fishery models can provide valuable input to ecosystem theory in at least two ways.

A large body of relatively accurate data exists for many commercially exploited species of fish. Landing records for Atlantic herring, upon whose bones, it has been said, the foundations of Amsterdam were built (Hardy, 1959), probably go back to the eighteenth century. In our country, landings from the enormously productive menhaden fishery are recorded back to 1873. Landings alone are not especially valuable data, but when a fishery is intensive and it can be assumed that fishing comprises a large proportion of the total mortality, then landings begin to give a reasonable notion of productivity of the system. When, as is frequently possible, we can couple landing data with effort data and age composition data; then, with the aid of the models described previously, we can more accurately estimate mortality rates due to fishing and due to all other causes. This provides a basis for more accurately scaling the ecosystem. For example, in figure 2, Steele (1974) has presented a production food web for the North Sea. While many of the intermediate

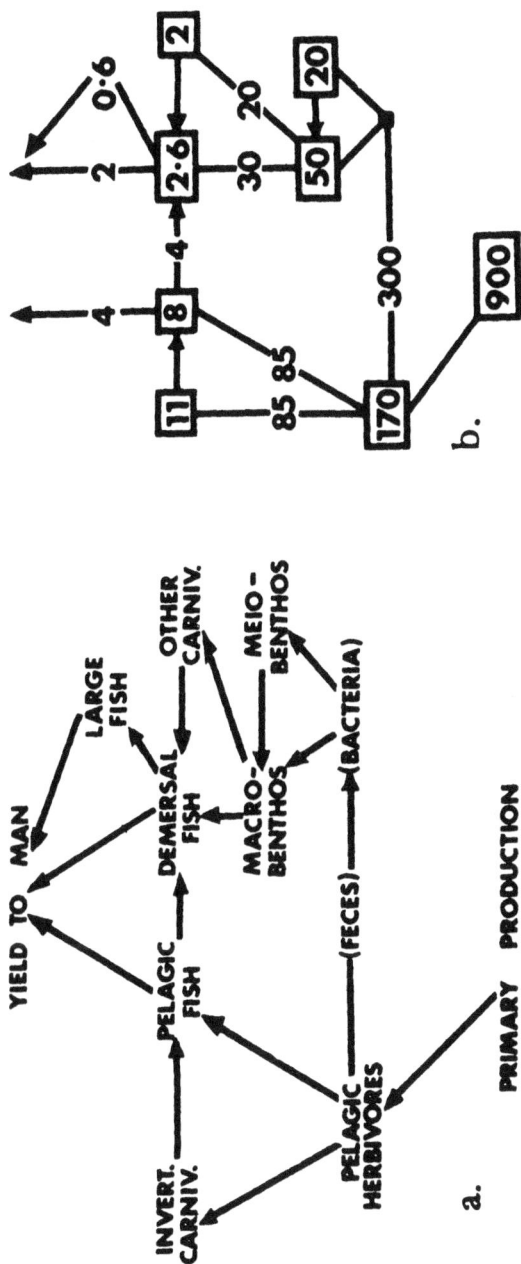

Figure 2. A North Sea food web based on the main groups of organisms. b. Values for yearly production (kcal/m² year) (from Steele, 1974).

values are speculative, based on analogy and simplifying assumptions, the whole system is bounded by fairly well determined values for the primary productivity and the fish yield. Even though the numbers are tentative, Steele (1974, p. 25) has stated, "They do indicate, again, that transfer efficiencies around 20 percent appear to be required of the pelagic herbivores and also, possibly, of the benthic infauna that feed on fecal material. The numbers could be arranged in various ways, but this would not alter one conclusion - that the yield of commercial fish is high in terms of the food web on which it is based." Steele (p. 14) concluded, "Thus data from commercial fisheries are invaluable in providing a minimum rate of output from a food web."

Steele has inferred, from looking at a system dominated by exploitation of carnivorous fishes, that overfishing likely affects only the exploited resource and not the structure of the whole system; that control of the marine ecosystem may be invested in the plant - herbivore interaction, rather than the herbivore - carnivore link as in terrestrial systems (Hairston, Smith, Slobodkin, 1960). What happens then when a herbivore is overfished?

At our laboratory we hope to construct soon a food web picture of this sort for menhaden. Based on fishery models and extensive data from the commercial fishery we have fairly good information of population size and input-output rates. Laboratory experiments are elucidating metabolic rates and growth efficiencies. The results might be quite interesting from the structural - stability point of view, since menhaden are heavily exploited and are herbivores.

Fishery input to stability of ecological models

Because fisheries, are predator-prey interactions, they provide data for examining the functional nature of feeding links. These functions may be the sources of ecosystem stability or instability (Steele, 1974; Smith, 1969, 1972; Rosenzweig, 1971; Hackney and Minns, 1974; Holling, 1966; May, 1973).

The traditional fishery models assume that the fishing mortality rate, F, is directly proportional to effort (i.e. $F(E) = QE$), so multiplying the catch rate by the biomass, the equilibrium catch per unit time would be $Q B E$, where Q is the proportionality constant. This is the Lotka – Volterra predation rate, which contributes neither stability nor instability to the system, as shown by Smith's (1972) analysis. Smith further showed that biological responses which increase the exponent of B, relative to E, increase the stability of the system, and vice verse. The Lotka-Volterra predation rate, and a logistic growth function for the prey species, result in a linear relation between catch per unit effort (CPUE, an index of population size) and effort (figure 3,A).

Developments in the fishery literature to non-linearize this relation (Pella and Tomlinson, 1969; Fox, 1970) are such as to stabilize the system by (implicitly) assuming Q varies proportionally to B - in effect increasing the exponent of B relative to E. As population decreases members of it become harder to catch, thus slowing the rate of decline. This would give a curve relating CPUE to E like that in figure 3,B.

The proportion of the population removed by one predator, or one unit of fishing gear, will increase with declining prey population,

CATCH/EFFORT

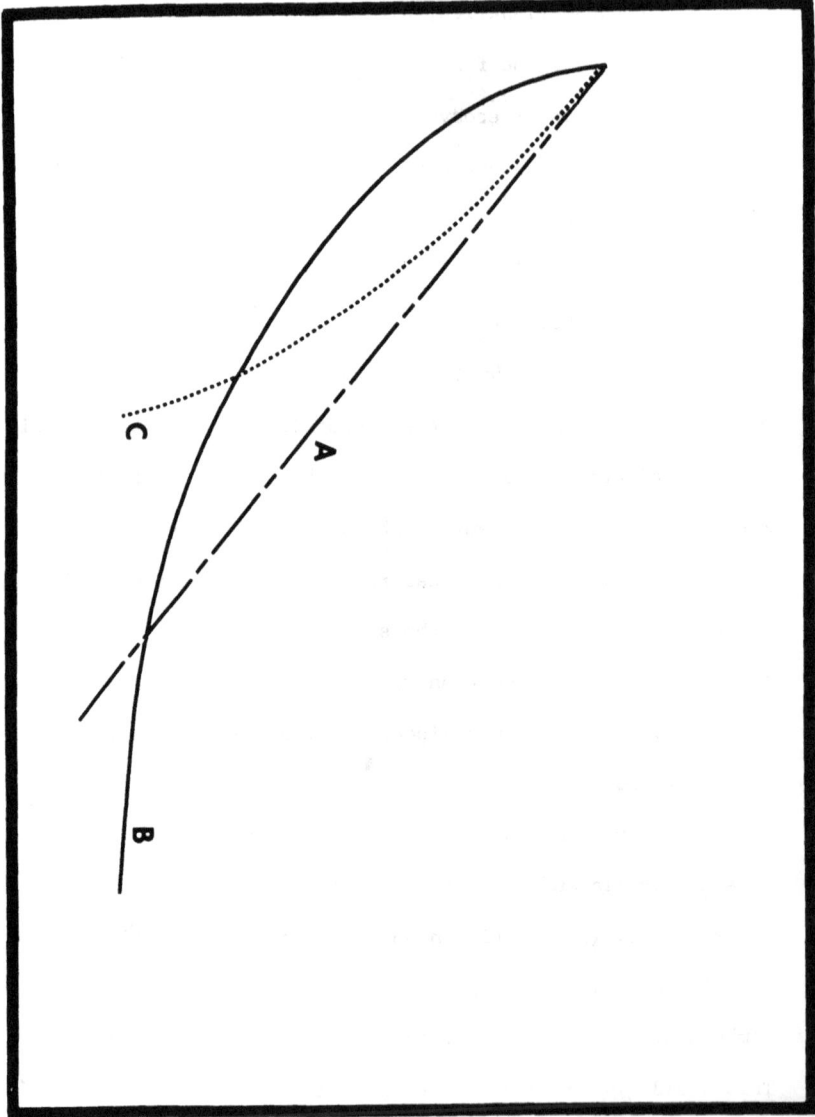

EFFORT

Figure 3. Three hypothetical relationships between catch per unit effort and effort.

if Q (the predator efficiency) varies inversely with B. The exponent of B decreases relative to E, and the system is less stable. The resulting CPUE - E Curve is that shown in figure 3,C.

The behavior of menhaden, which school densely and are visable at the ocean surface by spotter aircraft, result in this inverse relation between Q and B (figure 4). Part of the increase in Q may be attributable to technological improvements of the fishing gear (Nicholson, 1971). Never-the-less, exploitation in this fishery has a positive feedback on the exploitation rate. If we want to manage such systems, safe levels of harvest must account for this phenomenon, which has not been demonstrated for any fishery before. Fox (1970), in discussing fisheries with a type 3,B response, said, "This phenomenon indicates that it is not necessary to place such a tight restriction on the level of fishing as implied by the linear model to maintain an equilibrium yield close to maximum." The menhaden fishery, with its type 3,C response indicates just the opposite; tighter restrictions on the level of fishing are required to maintain maximum equilibrium yields, or even to maintain an equilibrium at all. If we want to understand persistant "natural" systems that behave this way (i.e. fast learning predators on clumped, visable prey), we must search for strong stabilizing factors.

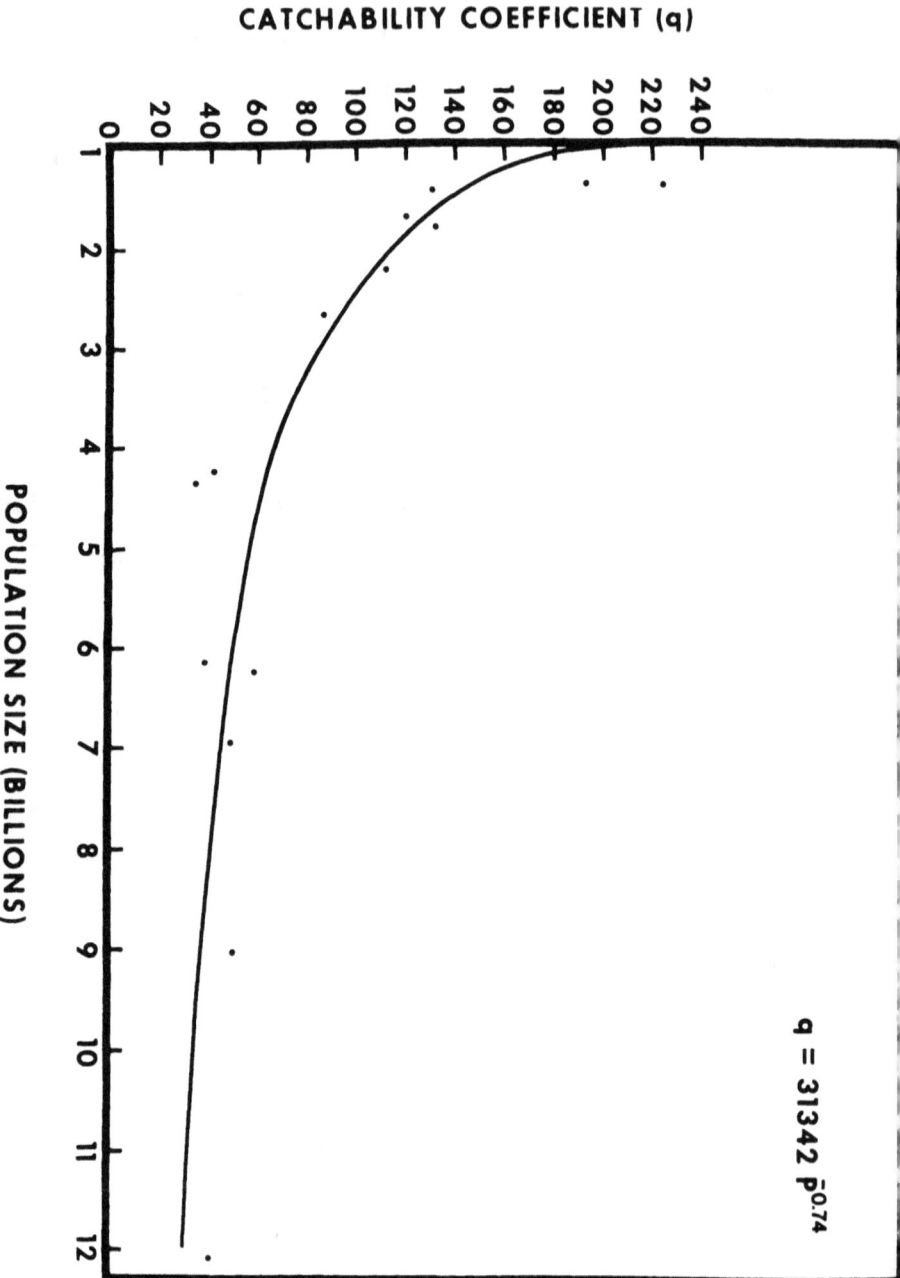

CATCHABILITY COEFFICIENT (q)

$q = 31342 \; \bar{p}^{0.74}$

POPULATION SIZE (BILLIONS)

Figure 4. The relationship of the catchability coefficient to population size, for Atlantic menhaden.

The impact of ecological modeling on fish population dynamics models

In spite of all the potential benefits of fishery data and models to ecology, I believe the really valuable interaction will be directed the other way. Ecological theory will have a profound influence on our views of how to manage fisheries, and how to build better models to guide our management policies. I believe this will come about primarily through a realization of the holistic view, that the structure of systems (at least mathematical models of systems) imposes a dynamic behavior on the components that is independent of the parameters used in the model (i.e. independent of what species are "plugged into" the model). As Smith (1969) states, "Suppose, for example, that one were interested in the dynamics of plants and made extensive measurements of their density, growth rate, metabolism, mineral adsorption, losses to grazing, and so on. The same plant in the same habitat, with the same total nutrient density, would show one set of characteristics if secondary predators were present and a very different set if they were absent. The investigator may be able to describe the state of his plants with precision, but unless he takes into account the whole system in which the plants participate, he cannot possibly understand why those results are obtained."

Before discussing further some of the implications of the systems view, I'd like to use the menhaden fishery again to illustrate a more prosaic application of ecological theory, though one that, to my knowledge, has not been demonstrated before. Fishery analysis has concentrated on the response of the target species to exploitation, but in

Schaefer's (1954) seminal paper he presented an equation from the
ecological literature (Lotka, 1925) for the growth rate of fishing
effort in response to changes in the fish population in an unregulated
fishery. The resulting pair of simultaneous differential equations

$$\frac{dB}{dt} = B K_1 (L-B) - QBE$$

$$\frac{dE}{dt} = K_2 E(B-b)$$

(3)

has no general analytical solution, but an approximate solution can be
obtained by numerical methods (Pielou, 1969), and Lotka (1923) analyzed
the nature of the solution to a pair of more general equations which includes (3) as a special case. From Lotka's analysis we expect the solution (a function of B and E) to be a spiral winding inward to a stable
equilibrium point. Schaaf and Huntsman (1972) presented a graph of
equilibrium catch on effective (adjusted) effort for data from 1955
through 1965. Effort was adjusted to hold Q constant, as required in
equations (3). That analysis provided the basis for estimating the necessary parameters for numerically approximating the solution to equations
(3). An updated version of their graph, shown in figure 5, extends the
expected trajectory of catch and effort through 1972. The dashed line
is not a statistical fit to the observed points, but is an extra-
polation of equations (3). The stable equilibrium point (1200 units
of effort yielding 330,000 metric tons of fish) provides a basis for
evaluating management strategies, such as a maximum sustainable yield
(MSY) point of 600,000 tons with 1000 units of effort. The benefits of
an MSY strategy should be evaluated, however, against the discount rate

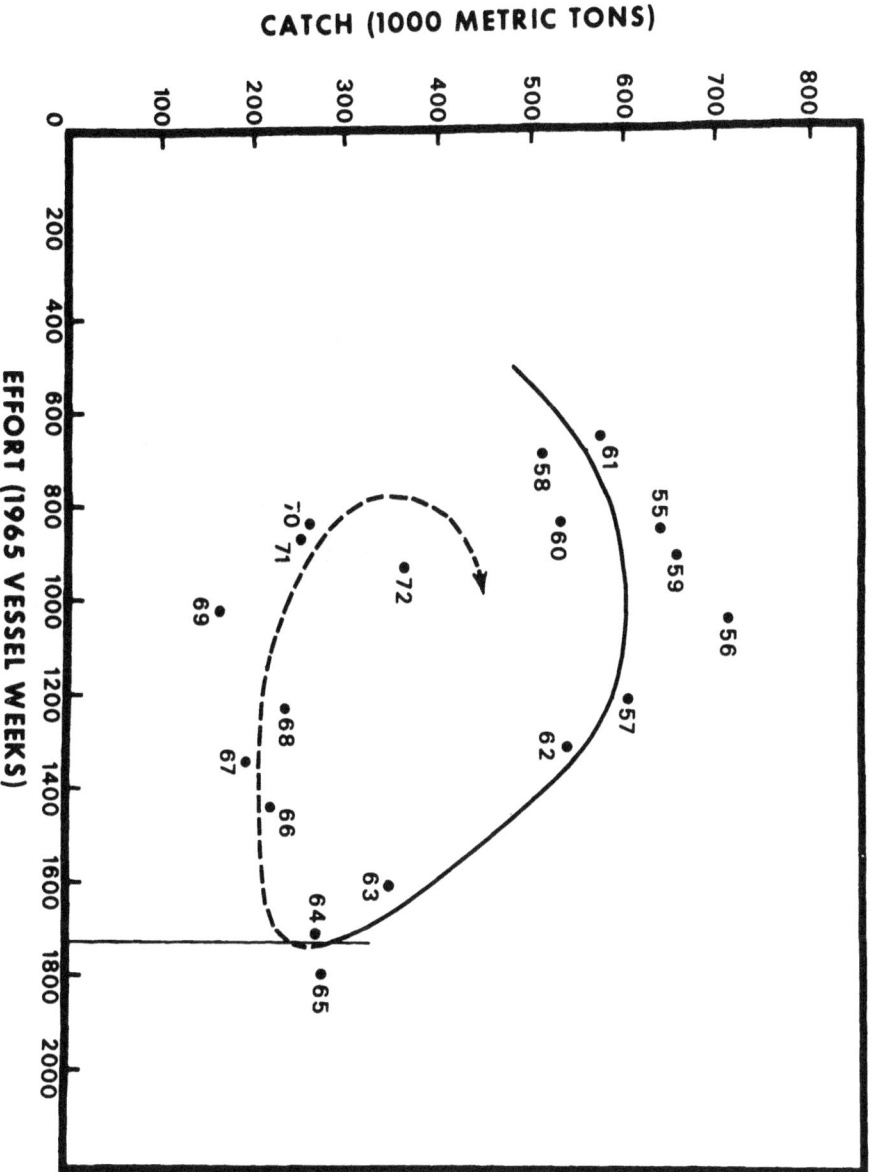

Figure 5. The relationship in the Atlantic menhaden fishery of equilibrium catch to effective effort (solid line) and the solution of Schaefer's simultaneous differential equations for an unregulated fishery (dashed line). The verticle line indicates the "biological break-even point."

of the industry and the time required to move from one equilibrium point to another. On the other hand, allowing the fishery to achieve its own equilibrium point carries risks, predicted from ecological theory (Watt, 1968) and observed many times to the dismay of fishermen dependent on a particular resource. The vertical line in figure 5 indicates a rough, first estimate (from simulation studies of a self-regenerating Beverton model) of the maximum exploitation rate that the menhaden population can withstand on a sustainable basis.

Aside from this kind of specific application of ecological theory, however, general trophic-dynamic models can provide us with a far broader viewpoint for considering the effects of exploitation, and rationalizing management. Some of our conclusions based on analysis of the exploitation of single species may be at best imprecise, and more likely, incorrect enough to be costly in terms of misdirected research, or consequent management.

If fishery systems are interacting subunits of larger ecosystems which have certain behavioral characteristics (e.g. constant production to biomass ratios) imposed by their structure, then we may be wasting our time getting very precise estimates of input-output rates, instead of estimating total biomass and position in the food web. Or, if length of food chain does not affect total production of a particular predator, but only its mean size (Kerr and Martin, 1970), then we should attempt to manipulate the length of the food chain in addition to any regulation of harvest rates.

Riffenburgh (1969) has presented one of the most clear-cut examples of how the application of trophic-dynamic theory can modify the results of a more simple fishery model. The fishery model implied that the yield of California sardines was quite sensitive to levels of fishing mortality. Analyzing a larger system model that included interactions with hake, a predator on sardines, and anchovies, a competitor of sardines, he concluded that sardines were not nearly as responsive to fishing rates; but sardines yields could be greatly increased by simultaneously harvesting anchovies and hake.

A more abstract example of the relation between models and predictions of yield from various fishing rates is given by Dickie (1973). A management strategy which regulates fishing effort to optimize yield efficiency would choose different levels of effort depending on which measure of yield efficiency is considered (yield per recruit, yield per unit food available, or yield per unit food consumed). The yield efficiency measure used depends on the type of food chain model employed.

Future perspective

I hope the preceeding discussion illustrates my belief that traditional fishery biology will benefit greatly from a broader ecosystem viewpoint, and that the extensive data and population modeling of fishery biology can play an important role in refining ecological models. Both approaches, however, may benefit from an even broader view of the world that considers and tries to account for multiple stable points, (Sutherland, 1974). I do not know what methodology will be

appropriate for the task, but in addition to building models that
faithfully represent the behavior of a system near its equilibrium
point, we need to focus attention on the boundaries of the system,
beyond which we no longer have the same system.

One of the more spectacular examples of this is the change in
the deepwater fish community of Lake Michigan, described by Smith (1968).
The outcome of a Lotka-Volterra competition model can be modified by man
(and other "new" predators) imposing additional mortality rates on the
species (Larkin, 1963); this apparently happened in Lake Michigan. From
1900 to 1945 lake trout supplied a fairly stable commercial fishery -
probably near the maximum sustained level. It collasped rather ab-
ruptly, to near extinction, by the early 1950's - by which time sea
lampreys, a recent invader of the lake, had established significant
numbers. The combination of high fishing rates and the additional
lamprey predation was enough to tip the balance. With the disap-
pearance of lake trout a succession of changes occurred among seven
competing species of chubs (<u>Leucichthys</u>) as the commercial fishery and
the sea lamprey switched to them - successively from larger to smaller
species. By the mid-1960's the deep water community was completely
dominated by the smallest species of chub and one other species, the
alewife, another recent invader. By the late '60's the alewife alone
dominated the system; and Smith conjectured that its production would
not equal that of all the once abundant species, since alewives do not
fully utilize the invertebrate food supply. He states that, "Greater
stability and full productivity of Lake Michigan...can only be achieved

by careful regulation of the kinds and numbers of predators, and the reestablishment of a multiple-species complex of prey species." Another of his conclusions is worthy of careful consideration – that greater stability occurs at low population density. Even if this is so, it may be that maintaining higher, and more fluctuating densities, will impart greater resiliency to the whole system (i.e. its ability to withstand additional perturbations without losing its structural configuration). A simple yield model of the lake trout fishery, or even a trophic-dynamic ecosystem model, constructed in 1950 would not have helped us predict this system response.

If, as Lewontin (1969), and Sutherland (1974) pointed out, history is important in determining the structure of communities, then we should be concerned with why a system is at a particular stable point. As I said earlier, I do not know how we do this analytically, but I believe we must be aware of the problem. We will be more cautious, at least, in applying our models to the management of natural resources.

In addition to the mathematical problems of studying system behaviour near boundaries, ecologists are faced with a practical pro-blem and a paradox that impede modeling ecosystems for resouce management-especially to predict effects of disturbance. Since our models are usually constructed to account for recent history, we need more exper-ience, of a structured type, with perturbed systems. We simply do not know enough about the effects of pollutants, say, on systems – or even on component species – to know how to incorporate them into our models. The paradox is that in order to understand ecosystems and predict their responses to perturbations we need to focus on emergent properties of systems; at present we do not know which of these properties will be most important to study; but whatever they are, they probably will not be of direct use to resource managers.

Literature Cited

Baranov, T. I. 1918. On the question of the biological basis of fisheries.
Izv. Nauchn. Issled. Ikliol. Inst. 1:71-128. (Transl. from
Russian by W. E. Ricker. 53 pp. mimeo.)

Beverton, R. J. H., and S. J. Holt. 1957. On the dynamics of exploited
fish populations. Fish. Invest. Minist. Agric. Fish Food (G. B.),
Ser. II, 9:1-533. H. M. Stationery Office.

Clark, G. L. 1946. Dynamics of production in a marine area. Ecol.
Monogr. 16, 323-335.

Crutchfield, J. A. 1973. Economic and political objectives in fishery
management. Trans. Am. Fish. Soc. 102:481-491.

Cushing, D. H. 1968. Fisheries biology, a study in population dynamics.
University of Wisconsin Press, Madison. 200 pp.

Deevey, E. S., Jr. 1947. Life tables for natural populations of
animals. Q. Rev. Biol. 22:283-314.

Dickie, L. M. 1973. Management of fisheries; ecological subsystems.
Trans. Am. Fish. Soc. 102-470-480.

Fox, W. W., Jr. 1970. An exponential surplus-yield model for optimizing
exploited fish populations. Trans. Am. Fish. Soc. 99:80-88.

Hackney, P. A., and C. K. Minns. 1974. A computer model of biomass
dynamics of food competition with implications for its use in
fishery management. Trans. Am. Fish. Soc. 103:215-225.

Hairston, N. G., F. E. Smith, and L. B. Slobodkin. 1960. Community
structure, population control, and competition. Am. Nat. 94:421-425.

Hairston, N. G. 1959. Species abundance and community organization. Ecology 40:404-416.

Hardy, A. C. 1956-59. The open sea, its natural history. Houghton Mifflin, Boston. 2 vols.

Harvey, H. W. 1950. On the production of living matter in the sea off Plymouth. J. Mar. Biol. Assoc. U. K. 29:97-138.

Holling, C. S. 1966. The functional response of invertebrate predators to prey density. Mem. Entomol. Soc. Can. 48:1-86.

Holling, C. S. 1973. Resilience and stability of ecological systems. Annu. Rev. Ecol. Syst. 4:1-23.

Jensen, A. L. 1971. The effect of increased mortality on the young in a population of brook trout, a theoretical analysis. Trans. Am. Fish. Soc. 100:456-459.

Kerr, S. R., and N. V. Martin. 1970. Trophic-dynamics of lake trout production systems. pp. 365-376. In: J. H. Steele, (ed.) Marine food chains. University of California Press, Berkeley.

Larkin, P. A. 1963. Interspecific competition and exploitation. J. Fish. Res. Board Can., 20(3), 1963. pp. 647-678.

Leslie, P. H. 1945. On the use of matrices in certain population mathematics. Biometrika 33:183-212.

Levins, R. 1968. Evolution in changing environments. Princeton University Press, Princeton, N.J. 120 pp.

Lewontin, R. 1969. The meaning of stability. Brookhaven Symp. Biol. 23:13-24.

Lotka, A. J. 1923. Contributions to quantitative parisitology. J. Wash. Acad. Sci. 13:152.

Lotka, A. J.   1925. On the true rate of natural increase of a population.
    J. Am. Stat. Assoc. 20:305.

Lindeman, R. L.   1942. The trophic-dynamic aspect of ecology. Ecology
    23:399-418.

MacArthur, R. H., and E. O. Wilson, 1967. The theory of island bio-
    geography. Princeton University Press, Princeton, N. J. 203 pp.

Mann, K. H. 1969. The dynamics of aquatic ecosystems. pp. 1-81. In:
    J. B. Cragg, (ed.), Advances in ecological research, vol. 6.
    Academic Press, New York.

May, R. M.   1973. Stability and complexity in model ecosystems.
    Princeton University Press, Princeton, N.J.   235 pp.

Nicholson, W. R. 1971. Changes in catch and effort in the Atlantic
    menhaden purse-seine fishery, 1940-68. U.S. Natl. Mar. Fish. Serv.,
    Fish. Bull. 69:765-781.

Odum, H. T. 1957. Trophic structure and productivity of Silver Springs,
    Florida, Ecol. Monogr. 27:55-112.

Pella, J. J., and P. K. Tomlinson. 1969. A generalized stock pro-
    duction model. Inter-Am. Trop. Tuna Comm. Bull. 13:421-496.

Pielou, E. C. 1969. An introduction to mathematical ecology. Wiley-
    Interscience, New York. 286 pp.

Ricker, W. E. 1944. Further notes on fishing mortality and effort.
    Copeia 1944:23-44.

Riffenburgh, R. H. 1969. A stochastic model of interpopulation dynamics
    in marine ecology. J. Fish. Res. Board Can. 26:2843-2880.

Ross, R. 1911. The Prevention of Malaria. 2nd ed. 679 pp.

Rosenzweig, M. 1971. Paradox of enrichment: destabilization of exploitation ecosystems in ecological time. Science 171:385-387.

Schaaf, W. E., and G. R. Huntsman. 1972. Effect of fishing on the Atlantic menhaden stock: 1955-1969. Trans. Am. Fish. Soc. 101: 290-297.

Schaefer, Milner B. 1954. Some aspects of the dynamics of populations important to the management of the commercial marine fisheries. Inter-Amer. Trop. Tuna Comm., Bull. 1(2): 27-56.

Schaefer, M. B., and R. J. H. Beverton. 1963. Fishery dynamics - their analysis and interpretations. pp. 464-483. In: M. N. Hill, (ed.), The sea, Vol. 2. Wiley, New York.

Smith, F. E. 1969. Effects of enrichment in mathematical models. pp. 631-645. In: Eutrophication: causes, consequences, correctives. National Academy of Sciences, Washington, D. C.

Smith, F. E. 1972. Spatial heterogeneity, stability, and diversity in ecosystems. Trans. Conn. Acad. Arts Sci. 44:307-335.

Smith, S. H. 1968. Species succession and commercial exploitation in the Great Lakes. J. Fish. Res. Board Can. 25:667-693.

Steele, J. H. 1974. The structure of marine ecosystems. Harvard University Press, Cambridge. 128 pp.

Sutherland, J. P. 1974. Multiple stable points in natural communities. Am. Natur. In Press.

Teal, J. M. 1957. Community metabolism in a temperate cold spring. Ecol. Monogr. 27:283-302.

Teal, J. M. 1962. Energy flow in the salt marsh ecosystem of Georgia. Ecology 43:614-624.

Thompson, W. F., and F. H. Bell. 1934. Biological statistics of the
Pacific halibut fishery. (2). The effect of changes in intensity
upon total yield and yield per unit of gear. Rept. Intern. Fish. Comm. 8,
44 pp.

Usher, M. B. 1972. Developments in the Leslie matrix model. pp. 29-60.
In: J. N. R. Jeffers (ed.), Mathematical models in ecology. Blackwell Scientific Publications, Oxford, 398 pp.

Van Dyne, G. M. 1969. The ecosystem concept in natural resource management.
Academic Press, New York. 383 pp.

Volterra, V. 1926. Variations and fluctuations in the number of individuals in animal species living together. Atti Accad. Nazl. Lincei,
Mem. Cl. Sci. Fish., Mat. Nat. 6(2):31-113. (English transl.,
R. N. Chapman, 1931, Animal ecology with special reference to insects,
pp. 409-448. McGraw-Hill, New York).

Wagner, F. H. 1969. Ecosystem concepts in fish and game management.
pp. 259-307. In: G. M. Van Dyne (ed.), The ecosystem concept in
natural resource management. Academic Press, New York.

Watt, K. E. F. 1968. Ecology and resource management. McGraw-Hill, New
York. 450 pp.

Williams, C. B. 1964. Patterns in the balance of nature and related
problems in quantitative ecology. Academic Press, New York. 324 pp.

Fisheries and Ecological Models in Fisheries
Resource Management

Robert T. Lackey

INTRODUCTION

The professional biases I incorporate into a review of fisheries
and ecological models in fisheries resource management are, in large
part, attributable to my orientation toward recreational fisheries
as found in North America. Freshwater fisheries scientists have
nearly always been more concerned with aquatic habitat and the whole
array of aquatic animal and plant populations than their marine
counterparts. The reason is quite understandable: the marine fisheries
scientist can rarely exert much influence on habitat or non-exploited
biota. On the other hand, freshwater systems may often be manipulated
as part of a management strategy. Both groups of scientists have been
quite concerned with target fish populations, and equally disinterested
in the third fisheries component, man (Lackey 1974a, Clark and Lackey
1975). The purpose of this article is to place fisheries and ecological
models into a renewable natural resource management context.

A good point to start an analysis of fisheries and ecological
models is by defining the system of concern: a fishery (either
recreational or commercial) is a system composed of habitat, aquatic
animal and plant populations (biota), and man (Figure 1). In a broad
sense, fisheries science is the study of the structure, dynamics,
and interactions of habitat, aquatic biota, and man, and the achievement of human goals and objectives through use of the aquatic resource.
Management is the analysis and implementation of decisions to meet
human goals and objectives through use of the aquatic resource
(Lackey 1974b).

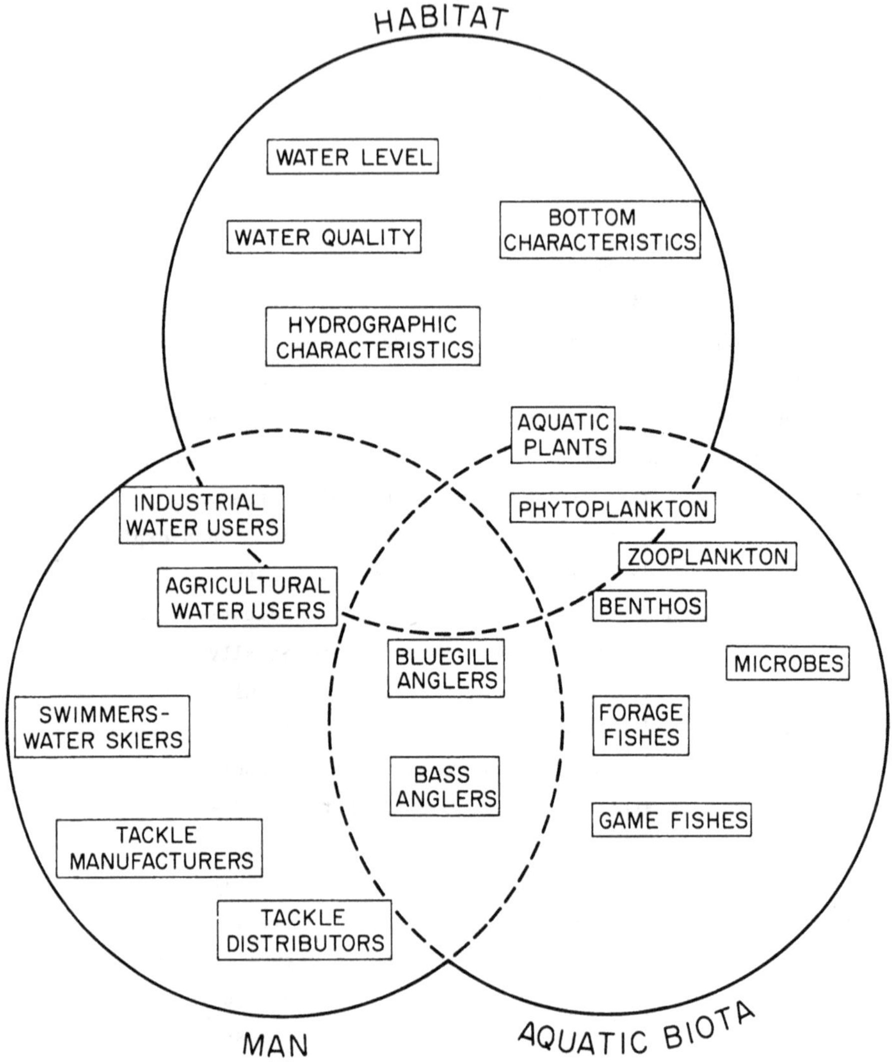

Figure 1. Schematic of a fishery system

Another concept needs to be clarified for the purpose of subsequent discussion: in a general sense, a _model_ is simply an abstraction of a system. Models may be verbal, graphical, physical, or mathematical (including computer simulation). However, renewable natural resources modeling nowadays usually connotes modeling of a mathematical nature. Throughout this paper, modeling will mainly be used interchangeably with mathematical modeling.

MODEL INTERRELATIONSHIPS

Most models, even those seemingly unrelated, are quite similar in philosophy and approach, but there is substantial variation between models when they are viewed according to their intended use or function. Models in fisheries management can be categorized into families that include one or more fisheries components (habitat, aquatic biota, and man) (Figure 2). The evolution of fisheries models has not followed a discrete path, but rather a disjointed and often circuitous route. The major trends (as exhibited in Figure 2) apply equally to recreational and commercial fisheries and marine or freshwater fisheries, but with different developmental paths being of greater importance.

Habitat models include those developed to predict aquatic temperature regimes, toxicant dispersal, and sediment transport (Figure 2). For example, one such management problem which exists in freshwater fisheries management is predicting the structure and function of proposed reservoir environments. Managers (and modelers) must first address and solve the problem of predicting future habitat characteristics, including physical and chemical parameters, before ecosystem and fisheries models can be consistently predictive. Predicting habitat characteristics is a difficult endeavor, but because it involves prediction of purely physico-chemical phenomena, it is _relatively_ easy.

Biological models include classical fish population dynamics models and models of single- and multiple-population systems. In this category we find the Schaefer and Beverton and Holt models (single population models in Figure 2). Nearly all of the extensive literature on population dynamics as applied in fisheries science

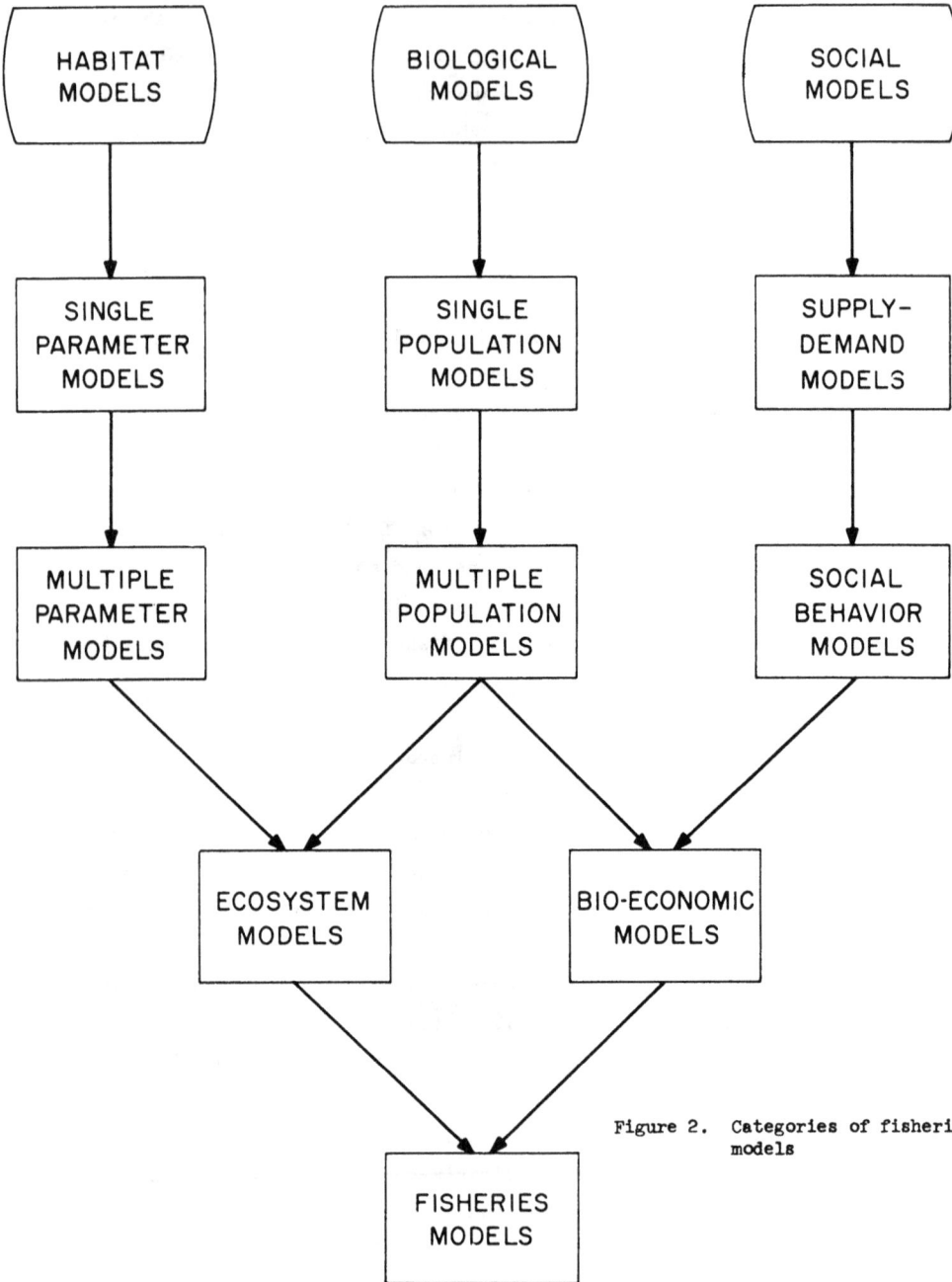

Figure 2. Categories of fisheries models

falls into this category. There has also been considerable activity
on developing biological models among ecologists (Smith 1974).

Ecologic or ecosystem models are becoming increasingly common
in fisheries science and other areas of renewable natural resources
management. Ecosystem models combine, in varying degrees, habitat
and biological models (Figure 2). Accounting for component interaction
is a key point in ecosystem models and much of the profuse literature
deals with interaction characteristics and mechanisms to describe them.
Freshwater systems have been modeled more frequently than marine
systems, in part due to the rather discrete nature of lakes and, to
a lesser extent, streams. The next step in ecosystem model development may well be an effort to solve the problem of managing an evolving
or unstable system.

Models that mainly address the third fisheries component, man,
fall into a category which may be termed social models (Figure 2).
In commercial fisheries, managers tend to measure fisheries output
as pounds of fish or perhaps net income. In recreational fisheries,
output is composed of many components, including aesthetics as well
catch (McFadden 1969, Lackey 1974b). From a management and modeling
standpoint, we must ask such questions as: How do men respond to
changes in renewable natural resources? How can human behavior be
predicted, or at least the behavior of part of the human population?

Bioeconomic models, as the name implies, include biological and
socioeconomic components of fisheries (Figure 2). Bioeconomic models
are integral to management of commercial fisheries (Pontecorvo 1973),
but neglected in recreational fisheries. Crutchfield (1973) has clearly
illustrated the role of social goals and fisheries management objectives.
Managing trends in use of aquatic renewable natural resources may
prove to be of much greater importance as human recreational and commercial
demands continue to increase.

Fisheries models, in the broadest sense, at least, combine
the major fisheries components (habitat, biota, and man) (Figure 2).
At such a comprehensive level of analysis, detailed modeling borders
on the impossible. However, if certain constraints (i.e. economic,
political, and social realities) are added to a comprehensive fisheries

model, one has a complete decision-making system.

POTENTIAL MODELING BENEFITS

Computer modeling in fisheries management can be justified in many ways, some of which result in benefit/cost ratios greater than unity and others that do not. As a group, fisheries modelers have, in my view, tended to oversell the potential benefits derivable from modeling, a characteristic all too frequent in emerging scientific disciplines. The potential benefits of modeling in fisheries management are many, and I would prefer to err on the conservative side as an advocate.

The first and perhaps most obvious potential benefit of computer-implemented modeling in fisheries management is organization. Fisheries are highly complex systems and modeling (graphical or mathematical) does provide a medium for clarification and organization. Used in this context, a model is a theory about the structure, dynamics, and function of a fishery or a fishery component.

A second potential benefit of modeling in fisheries management is a self-teaching device to the builder or user. There may be no better way to develop a "feel" for a fishery than to formally model it. Some fisheries models, particularly computer-implemented models, serve as useful management exercises in universities (Titlow and Lackey 1974).

Identifying gaps in our understanding of a system is a third potential benefit from modeling in fisheries mangement. In modeling, the modeler may become painfully aware of areas of missing data. Acquisition of these data may well be top priority for improving management. Sensitivity analysis in modeling will identify the parameters of most importance in determining model output, and data acquisition and/or research efforts may be allocated accordingly.

Models as research tools may be considered as a category of potential benefits. Manipulation of the model itself may generate "data" which is unattainable from the real system. For example, the impact of rainfall and water temperature may each have an impact on certain biotic components, and certain combinations of rainfall and temperature levels have been observed in the field to quantify the impact. Exercising the model may permit a reasonable assessment of the general relationship by interpolation (based on existing data combinations).

The fifth and most discussed potential benefit of modeling in fisheries management is predicting the impact of alternative management decisions or external influences. Historically, managers of commercial fisheries have been interested in predicting the impact of a proposed fishing or exploitation rate expressed in the form of a season, mesh size, or quota. Recreational fisheries managers wish to estimate the impact of decisions on the number of realized angler-days, catch, or some other measure (Lackey 1974b). As a very general guide, habitat models will potentially possess relatively good predictive power, biotic models intermediate predictive power, and social models relatively poor predictive power.

MODELS AND DECISION-MAKING

The potential benefits of mathematical modeling are not universally accepted among professional fisheries scientists. Agencies supporting or proposing to support fisheries modeling will, in my view, increasingly demand a clear itemization of the expected benefits of modeling.

Fisheries management is a very pragmatic discipline and the results
of research efforts are generally expected to improve management
decisions. All too often researchers have failed to bridge the gap
between their work and the decision-making process. This is not to
say that we need a public relations campaign to advocate modeling,
but rather to present the research results in a useable manner.
Research is only one input in the decision-making process and its use
depends in part on ease of use.

As a final note about fisheries models and modeling, I foresee
a much closer involvement between "modelers" and "decision-makers."
The distinction between the two groups is purely artificial, but tends
to develop by a "division of labor" approach in structuring an agency.
Frequently, those actually making or recommending management decisions
perceive, at least subconsciously, modelers as a threat, or worse,
a pack of contemporary Don Quixotes. Modeling offers too much to
resource management to fall into this image.

Literature Cited

Clark, Richard D., Jr., and Robert T. Lackey. 1975. Managing
    trends in angler consumption in freshwater recreational fisheries.
    Proc. Southeastern Assoc. Game and Fish Comm. 28:(In Press).

Crutchfield, James A. 1973. Economic and political objectives in
    fishery management. Trans. Amer. Fish. Soc. 102(2):481-491.

Lackey, Robert T. 1974a. Priority research in fisheries management.
    Wildlife Society Bulletin 2(2):63-66.

_____, 1974b. Introductory fisheries science. Sea Grant,
    VPI & SU Extension Division, Publ. VPI-SG-74-02, 275 pp.

McFadden, James T. 1969. Trends in freshwater sport fisheries
    of North America. Trans. Amer. Fish. Soc. 98(1):136-150.

Pontecorvo, Giulio. 1973. Ocean fishery management discussions
    and research. NOAA Tech.Rept. NMFS CIRC-371, 173 pp.

Smith, J. Maynard. 1974. Models in ecology. Cambridge University
    Press, London, 146 pp.

Titlow, Franklin B., and Robert T. Lackey. 1974. DAM: a computer-
    implemented water resource teaching game. Trans. Amer. Fish.
    Soc. 103(3):601-609.

Management of Large-Scale Environmental
Modeling Projects

R. V. O'Neill

INTRODUCTION

Large-scale, interdisciplinary systems analysis programs are
relatively new to the environmental sciences. Management of any
scientific research program is complex, but integrated research
directed toward specific objectives adds new dimensions to the
problem. Additional technical difficulties arise when simulation
models play a key role in project integration and synthesis. This
paper addresses a single aspect of the total problem, management of
the modeling component of the program. Although, of necessity, some
comments will deal with the general problem, specific emphasis will
be placed on the models, the modelers and interactions between systems
analysis and other aspects of the program.

The analysis is based on the Eastern Deciduous Forest Biome, US-
IBP, and on my personal experience as its modeling coordinator since
1969. Since I know of no reliable method to filter personal bias from
the presentation, the reader must determine which portions of the analysis are relevant to his own situation. However, to assist the reader,
I recommend two recent reports (Van Dyne 1972, Mar and Newell 1973)
which represent other analyses of interdisciplinary environmental
modeling studies. By comparing these with the present discussion, some
generalizations may emerge.

Organizations, personnel and objectives differ; and each research
program may experience a different set of problems. However, as programs become more massive, something akin to the law of large numbers
seems to apply. A wider spectrum of individuals becomes involved,

and problems tend to converge. Therefore, despite the individuality
of my experiences, some of the analysis may prove useful to others.

The presentation is organized into four subject areas. First,
I will briefly discuss the Eastern Deciduous Forest Biome to provide
a context within which the reader can evaluate the validity of extra-
polating the analyses to other programs. Second, the nature of a
model and the modeling process should be understood since some characteristics apply to any program which relies on the synthetic capability of models. Third, scientists are people, and an explicit
consideration of the motivations and conflicts of the modeler suggest
criteria for the management structure. Finally, in addition to managing
the modelers themselves, project leadership must address questions
concerning the interactions between modelers and other scientists in
the program.

AN INTEGRATED ENVIRONMENTAL SYSTEMS ANALYSIS PROGRAM

The International Biological Program (IBP) was established in 1964
as a world-wide effort to study the biological productivity of the
earth with particular reference to human welfare. The Eastern Deciduous
Forest Biome (EDFB) began in 1969 as one segment of the United States
contribution to the IBP, which also included tundra, grasslands, conif-
erous forest and desert biome programs. These were all designed as
integrated research efforts, with interdisciplinary teams structured
to promote coordination, rapid communication and synthesis.

From an original planning proposal (Auerbach 1969), the EDFB has
grown to involve more than a hundred scientists at any one time and a

dozen universities and agencies (Burgess & O'Neill 1974). The objectives
of the program are numerous and quite broad, but all involve the development of an understanding of ecosystems at several levels of resolution
ranging from forest stands to regional landscapes. The broad goals
indicate a desire to begin a comprehensive assault on understanding ecosystems. The project has stressed detailed study of ecological processes
controlling terrestrial and aquatic ecosystems within forested watersheds,
the basic units of study.

At the beginning of the EDFB, much of the American ecological research
community was centered in universities in the Eastern U.S. Some of these
universities had already developed ecosystem programs with interdisciplinary teams. To maximize participation of the available talent, the program
was organized around five, semi-autonomous sites, chosen on the basis of
available talent in pertinent research areas, a demonstrated interest in
interdisciplinary research and the availability of modeling personnel.
The five participating sites were: Coweeta Hydrologic Laboratory of the
U.S. Forest Service in western North Carolina, administered through the
University of Georgia; Lake George, New York, administered at Rensselaer
Polytechnic Institute; Lake Wingra, headquartered at the University of
Wisconsin; Triangle site, in the North Carolina Piedmont, directed from
Duke University; and Oak Ridge, utilizing the AEC Reservation at the
Oak Ridge National Laboratory in eastern Tennessee.

Central to the research strategy has been the use of mathematical
models to integrate and synthesize data generated at the sites (O'Neill
1974). A group of modelers has been an integral part of the research
team at each site. These groups have had responsibility for models of

ecosystem processes being investigated at their respective sites, assisted by a central modeling group at EDFB headquarters in Oak Ridge. This central group has provided expertise in numerical analysis and computer sciences, has assisted in the development of some models for specific processes and has been responsible for the synthesis of total ecosystem models. Throughout most of the program there have been 10-20 modelers operating on various aspects of the project. The modelers were recruited from a variety of disciplinary backgrounds including mathematics, physics, engineering, biology, and chemistry and represented a wide array of viewpoints, training, and experience.

This terse description of the EDFB program should suffice to convey the impression of a large, complex environmental project with considerable emphasis on integrated research. It is hoped that even so brief a description will help place into perspective the management experience analyzed below. We will be focusing on interinstitutional, interdisciplinary programs with more than 50 participants and with a strong reliance on modeling to synthesize research results. Because of the nature of the EDFB program, the analyses may be most relevant to geographically diffuse projects that rely heavily on scientists housed at universities to supply the modeling talent. It will be apparent to an experienced manager that many of the problems discussed are less likely to occur in small, in-house modeling teams that have already achieved a degree of integration.

THE NATURE OF A MODEL AND ATTENDENT PROBLEMS

Some of the problems inherent in the management of a large-scale modeling effort result from the nature of models and the associated modeling process. Explicit consideration of several points can be invaluable in understanding these problems and in suggesting reasonable resolutions.

Model Objectives

By its very nature, a model is an analogy, a partial representation of a real system. A given system can be abstracted in any number of ways, and a near infinite variety of models, from very simple to extremely complex, can be hypothesized. The most useful criterion for choosing one particular model is the objective for which the model is intended. The purpose will determine the assumptions accepted and the viewpoints emphasized. There is no such thing as a truly general or universal model. There will always be other models with different emphases and resolutions, equally valid for different purposes. Models are not true or false; they merely possess different degrees of usefulness. They are accepted as valid only if the scientific community discovers that adoption of this specific abstraction of the world, this specific Gestalt, is helpful in solving a particular class of problems.

These observations may seem trite and obvious, but experience indicates that underestimation of the importance of objectives is a prime ingredient in the demise of modeling projects. Since abstractions and assumptions _must_ be made, specific objectives _will_ be set, either explicitly in project goals or implicitly by the modeler. If a model must answer specific questions, the questions should be established for the beginning of the project, not at the end. As the possible number of hypothetical models approaches infinity, the probability that any given model will do exactly what the manager wants approaches zero (Brewer 1973).

As a simple illustration, consider the problem of developing a "water resources model," and consider the ambiguities associated with the objective: "predicting the effect of perturbation on aquatic biota". This is not an operationally useful objective since it does not resolve the following dichotomies: Is the interest in phytoplankton biomass (DiToro et al. 1970), or simply blue-green algae without reference to biomass (Bartell et al. 1973)? Should one consider physical water movements (Chen and Orlob 1972) or ignore them (MacCormick et al. 1972)? Will emphasis be on nutrient dynamics (Walsh 1972) or on biological interactions (Park et al. 1974)? Is fish biomass sufficient (Kitchell et al. 1974), or must one consider species replacement (Baumann et al. 1974)? Should one be concerned with temperature (Water Resources 1968) or urban development (Huff et al. 1973)? The resolution of these questions can lead to very different models, as examination of the above references demonstrates.

It is easy to surmise that sufficient specifications have been set on modeling objectives at the beginning of the project, and to learn later that communication gaps have resulted in a relatively useless product. The specific objectives, therefore, must be stated with great precision. A potentially useful method lists specific input data and

specific output parameters which can be used as environmental indices.
A model essentially processes a set of input numbers to produce a set
of output numbers. Whether or not a model is relevant to a problem
depends on whether or not the model will accept the available input
data and produce output that corresponds to the desired output set.
This approach is well known to the water quality engineer who has long
developed models to predict specific parameters of system responses,
e.g., dissolved oxygen, BOD, Coliform MPN, etc. Unfortunately, such
specific indicators have not been developed for more general considerations of the effects of perturbation on aquatic biota. The manager
of a modeling effort should clearly list the input information available
and specific indicators desired as output. While this in no way guar-
antees that the output will be correct or processed in the best possible
manner, it will at least help ensure that the model is aimed at the
correct target.

The Modeling Process

A complex model, particularly for a poorly understood system,
never seems to be completed. At any point in time, the model paradigm
synthesizes our understanding of certain aspects of the system's behavior.
If the system being modeled is poorly understood, the model may contain a
wealth of hypotheses about interactions and factors which affect a specific
output set. As research projects are completed, some of these hypotheses
will be rejected and new functional forms will be suggested. For example,
earlier models of ecosystems frequently postulated that consumption or
grazing rates were proportional to the product of the magnitudes of food
source and consumer (Lotka 1925, Volterra 1928). But continued research

(Smith 1952, Minorsky 1962) now indicates the need for more elaborate expressions (Watt 1959, Holling 1959, Ivlev 1961, DeAngelis et al. 1973, Wiegert 1973) since direct proportionality to the product probably exists only over a limited range of intermediate values. The point is that new research results are continuously increasing our insights into ecosystem function and the models develop as an almost continuous process through time.

Complex models are usually implemented as computer codes. The larger and more complex the model, the greater seem the possibilities for improving the code and its efficiency. There is always one more item that needs refinement and, unfortunately, each alteration seems to reveal more about the model and to indicate need for further improvements. In other words, as Brewer (1973) points out, if the modeler waits to "finish" the model to his "satisfaction," the product will be significantly delayed. A reasonable resolution is to enforce periodic documentation of successive versions of a model during its development.

Model Transferability

Modeling can be a highly individualized art form. At present, the skill, ingenuity and experience of a modeler are required to translate scientific insights into mathematical formulations. Computer programming in higher level languages such as FORTRAN can also become highly individualized. The experienced programmer may evolve his own "bag of tricks" which he finds useful and effective. As a model develops, different sections may be coded by different individuals, and modifications may be introduced. There are also significant machine-dependent problems, particularly those associated with the handling of large data sets (e.g., availability of drum vs. disc vs. tape files). As a result, the model cannot be easily transferred to others. The original modeler understands it and can utilize it, but it may be a tremendous problem for someone else to understand the model well enough to make use of it. This problem is not unique to environmental programs (Brewer 1973).

An interesting case history of the problems associated with model transfer can be seen in the hydrologic transport model developed in our program (Huff et al. 1973). The model began with the Stanford Watershed Model (Crawford and Linsley 1966) which was modified to include the transport of materials (Huff 1968). The model then moved to the University of Wisconsin where it became an important element in the EDFB Lake Wingra project (Huff et al. 1970). The move required a translation of the program from PL/1 into FORTRAN and other changes to permit implementation on a UNIVAC computer with limited data handling capabilities. The program, now called the Hydrologic Transport Model (HTM), was documented in 1972 (Huff), and work began on incorporating a series of improvements, including open channel hydraulics (Jacques and Huff 1972a), impoundment effects (Ivarson et al. 1972), snow melt (Jacques and Huff 1972b), soil infiltration (Reeves and Miller 1974), and transpiration (Goldstein and Mankin 1972). By this time the model had been touched by a considerable number of workers and had become increasingly complex. As a result, six months were required to transfer the model to another program (Fulkerson et al. 1974) to simulate the transport of trace contaminants.

Because a model is individualistic and its development may be proprietary, the problem may be viewed as inherent in the modeling process itself. The model may be relevant and scientifically elegant and still be useless to the community at large. Therefore, if the model is intended for general use by others, this goal should be explicit from the beginning. The project must permit some independent organizations (agencies, universities, consulting firms) to implement the model on their own computers. If this objective is clear from the beginning, documentation may be more adequate and the model is more likely to be free of idiosyncrasies.

Model Validation

A model for a complex environmental system can be assumed to be invalid with a fair degree of confidence (Passioura 1973). To say that it is valid is to claim a basic understanding of ecosystems which does not exist. That is, to maintain that the model will make quantitatively correct predictions, implies an understanding of the intricate control mechanisms and feedbacks which regulate the system. The science of ecosystem analysis is far too immature to support such a claim. Present models do not adequately deal with many ecosystem processes which are essential in accurate predictions. For example, specific perturbations may not lead to simple changes of rate processes, as presently modeled, but to species replacement instead. Present models (e.g., Kitchell et al. 1974) may predict reductions in fish biomass, but actual system response may involve replacement of present populations with less desirable species (Baumann et al. 1974).

On the other hand, if the model has been developed by competent
workers, it is also valid, since one can be reasonably sure that it
will match those aspects of system behavior which were emphasized in
its development (Levins 1966). Thus, it is quite possible to invalidate or validate a model by manipulation of questions asked. It is
easy to demonstrate that the Iliad is an atrocious novel; it is also
not too difficult to demonstrate that it is the best Greek epic poem
written by Homer concerning the fall of Troy.

Despite the ambiguities associated with validation, some confidence
in model reliability and limitations must be established before the model
can be considered useful in management applications. Therefore, tests
of its capabilities should be conducted. But further testing by the
original developer reaches a point of diminishing returns. Extensive
(and possibly expensive) tests should be conducted by an independent agent,
who can approach the problem objectively and from a new vantage point.
The desirability of implementing the model on another computer was discussed earlier as a means to minimize the problems of transferability.
The independent agent can serve both functions. The independent testing
is not designed to question the integrity of the modeler but to examine
the strengths and weaknesses of the model from a different viewpoint.
This process is similar to the duplication of experiments in a different
laboratory.

MANAGING THE SCIENTIST-MODELER

Modelers are people, and herein lies the essence of the management
challenge. The mechanisms and motivational structures established for

the project are key factors in its success or failure. Management is an art not readily amenable to the type of analysis which is attempted here. Much of what is outlined will seem trivial to the experienced and perhaps paranoid to the novice. Suffice it to say that there are management problems characteristic of geographically dispersed, university based modeling projects which must be understood and addressed. I will not attempt to address all aspects of scientific management, but some general overview of the characteristics of the scientist will facilitate my discussion of managing the modeler.

The Scientist and His Motivational Structure

Scientists are people and have the basic motivational structure common to all human beings. Security, personal development and ego gratification are important considerations. As a result, management must recognize that a scientist is not involved in a project solely for the good of the project. His motivational structure is a complex mixture of relatively mundane considerations as well as idealism and intellectual challenge.

But scientists are also intelligent people and techniques which are effective in other management contexts will not necessarily assure good performance. Obviously, to achieve project goals, direct control may sometimes be necessary, for example, by withdrawing funding support. At other times indirect control through peer pressure may be required to produce a specific product. But in inter-institutional programs, the project manager is unlikely to be a direct supervisor and the scope of available control mechanisms will be relatively limited. The most creative scientific endeavors can be expected to occur in an environment in which the project is viewed by the individual scientist as a professional opportunity. Many management techniques can be evoked to produce a product, but when the product requires creativity and direct control is limited, the situation is more complex. In my experience, the most effective approach calls the scientist's basic motivational structure into play so that he attempts to produce high-quality science without having to be manipulated.

The situation is still more complex since the best scientists (who will form the core of any program) will most likely be exceptionally intelligent and creative individuals. They may dislike deadlines, regimes, constrictive bureaucracies, orders and programming. They may be erratic and impulsive with tremendous periodic drive. At some times and for some people, teamwork frustrates this creativity and some scientists never feel comfortable in this environment.

The difficulties of integrated research are also complicated by reward systems that conflict in the demands placed upon the individual. A majority of the scientists involved in the project may be housed in traditional university departments. The University department may impose a reward system in conflict with the reward system established for the integrated project. The department may impose demands for teaching, committee work, etc. which conflict with project demands for research. The departmental motivational context emphasizes the solitary worker, individual initiative and creativity, and stresses work in a restricted disciplinary field (Van Dyne 1972). Some changes are already evident since major funding of interdisciplinary studies has encouraged changes of policy, but this motivational conflict cannot be ignored. Stated quite bluntly, the structure encourages the individual to present himself as totally cooperative and wholeheartedly dedicated to the goals of the project in order to obtain research funding, and then to utilize the funding in a strictly disciplinary study in isolation from the project. The problem is a difficult one and its resolution depends upon continuing changes in the traditional motivational structure of the university.

The preceding discussion develops an image of the scientist as strongly individualistic, impulsive, egotistical, and with conflicting motivations. But while the individual scientist may tend toward these characteristics, I do not wish to imply that the emotionally mature, cooperative scientist does not exist! He most certainly does. In fact, our program has been staffed with a number of strong-willed but extremely cooperative scientists who have formed a stable and productive base throughout the project. Strauss and Sayles (1972) point out that although the professional may be erratic in his work habits and desire strong independence from project organization, the conflicts can easily be exaggerated. The professional may require special treatment, but his self-motivation makes him a hard worker. He will frequently relate to project goals as intellectual challenges and may indeed require little or no management encouragement. Nonetheless, as the characterization given above is exemplified by individual personalities, one sees the unique management problems associated with the creative scientist. And it is these unique problems, rather than the total human management problem, that we wish to address here.

Management of the Modeling Project

Let us proceed to apply this image of the scientist and the concepts of motivational structures to the actual problem of managing the modeling aspects of the project. Both management and modeler have specific roles to play, and the interactions and demands placed by one upon the other are important to the successful completion of the project. Explicit consideration of the motivations of the scientist can assist in establishing these interactions in a manner conducive to project success.

To accomplish program goals, the project requires leadership, but if project management develops an image of the "grasping Capitol city with its subsidiary satellites", revolution is the logical and historically inevitable conclusion. And yet total democracy is inefficient and counterproductive since everyone contests for superiority. Whether or not the program is designed around a central headquarters, an image of autocracy is to be avoided. This image is enhanced by actions which seem to distract or interfere with the science, and the image is reversed when decisions seem to provide new opportunities.

An excellent example of applying the explicit motivational structures of the university scientist to management of the modeling project can be seen in the development of the biological components of our lake ecosystem model, CLEAN (Park et al. 1974). The development of the total model required creative input from a large number of individual biologists. To accomplish the task within the proper motivational context, the biological component of the total model was divided into a number of modules. Each module was developed by an individual scientist working with a central modeling team from project headquarters. The central team worked to ensure that the modules were compatible and addressed the objectives of the total model, but the individual modules were clearly identified as the intellectual product of the individual scientist. It was his model and was implemented on his computer for purposes of his own investigations. He could identify the model as his own and could see the professional and intellectual advantages to producing the best possible model. Each module served to assist in synthesizing his own data and was available to investigate the scientific questions of specific interest to him. Thus, the objectives of the program were achieved in developing the biological components of the total model, while the individual scientist was able to satisfy his own motivational structures and was encouraged to exert his best creative talents.

An important element in establishing a proper image of leadership is the mode of communication. This is especially important if the modeling project is geographically dispersed. Minimize paper work. Whenever possible, personal communication will permit the modeler to comprehend a decision in a manner that is more satisfying to his ego and personal intelligence. When efficiency and project demands require written communication, some care should be exercised in execution. For example, every written communication should be composed of two parts: The first should begin, "Please...", and the second, "Because...". Given the intelligence of the modeler, he can supply his own third part beginning, "Or else...".

Drucker (1974) discusses some of the problems associated with the management of the career professional. Management plays a role in placing the professional's ideas into the context of a total program, but in a very real sense, the professional is the "superior". He is the generator of the insights, the true director. He must participate in program direction if he is to be more than a mere technician. Management serves more to provide an adequate environment than to dictate daily activity. Therefore, in many respects, management must view itself as a support activity.

The lack of appreciation for careful communications can be exemplified by the following kiss of death: "This decision is so clearly logical that no one could possibly misinterpret my reasons for making it." The modeler will cooperate, but his analytical mind will want to know the reasons for a decision and why it is to his advantage and the advantage of the project. If the management is actively endeavoring to develop the proper motivational environment for creative research, the modeler should feel that management is performing a necessary function for him.

At the beginning of the EDFB, a number of research groups had already begun the development of models for the forested watershed. No single group appeared to have the clear and unique insight to undertake the total modeling task. Therefore, deciding to adopt the viewpoint and emphasis of one group rather than another would have worked against the creative impulses and intellectual drive of the total program. The project management decided to provide considerable latitude in model development, even at the risk of losing efficiency. The results seem to justify the decision to develop an environment for creativity rather than impose constraints on the modeling projects. Indeed, the broad range of questions being asked in the program probably could not have been addressed by any single approach to modeling. Questions relevant to ecosystem dynamics appear at the scale of the responses of single leaves to changes in microenvironment (Murphy and Knoerr 1970); other questions require a scale of weeks or months and consider responses by the total plant (Sollins et al. 1974, Shugart et al. 1974); still other problems consider total landscape units and changes in community structure occurring over periods of many years (Goldstein and Harris 1973). We now understand more adequately that the broad problems of ecosystem analysis may best be approached by a simultaneous attack at several levels of time and space resolution (Goldstein and Mankin 1971). And no single modeling approach which might have been adopted originally on the grounds of efficiency would have taken us as far toward a basic understanding of ecosystem dynamics.

In addition to the general image of leadership conveyed by management, particularly in methods of communication, some operational principles follow logically from consideration of motivational structure, for example, the choice of participants in the program. The modelers with the best record of productive, creative research are most likely to accomplish the goals of the project. During the first four years of our program, there has been a core group of highly productive workers. A dozen individuals have contributed to over half of our open literature publications produced during this period.

While individuals with a good record of productivity will play
an important part in achieving goals, real difficulties are involved
in the creative aspects of program objectives. In any scientific
research program that involves more than application of known facts,
creative input is an important ingredient. And one of the fundamental
challenges of research management is the development of an environment
within which such creativity can flourish. Creativity and initiative
in the modeling effort must be encouraged but channeled to meet deadlines and produce specific output. Unfortunately, the truly creative
personality tends to be impulsive, egocentric and self-motivated.
Therefore, the creative individual may be the most difficult to manage.
However, the proper conclusion is not to select only mediocre scientists
because they are easier to manage; their models may also be mediocre.

A properly designed modeling project should therefore include
a balanced variety of studies that take advantage of individual
talents. Invest in some high-risk creative research with potential
for scientific breakthroughs; these projects may redefine the whole
program. Invest the largest portion on individuals with a history
of cooperation and productivity. Then, it is important to identify
clearly the portions of program objectives which can be addressed
by technically competent scientists who are expected only to produce
some specific, and predetermined, product.

Modelers often march to their own drumbeat. Sometimes the
beat is synchronous with project goals; sometimes the rhythm must
be checked and disciplined; occasionally the drumsticks must be
taken away. Therefore, long range funding commitments should not

be made at the beginning of the project. Integrated programs evolve, and increasing support for productive researchers will result in reduced funding for the less productive. It is essential to have an internal review process, with objective criteria for judging productivity, and appropriate sanctions that can be utilized to encourage the productive and discourage the uncooperative.

Alternatively, positive motivations can be provided by opening other rewards to the modeling personnel. Encourage better scientists to address congressmen, travel overseas, etc. Share the prestige rewards that are frequently considered the exclusive domain of project management. By broadening the motivational base, the job is made simpler and the probability of success is increased.

Modelers Within the Management Structure

I have been attempting to examine modeling project management in a way that considers motivational structures and acknowledges the individual talents of the scientist. But the analysis also indicates that one should not develop a management structure and then go out and staff it. Instead, find the people and build the structure to suit their talents. Optimization of the individual talents and motivations of managerial staff is the subject of an interesting article by Ludwig (1974). He discusses the "Lifo" system of managerial styles devised by Atkins-Katcher Associates, Inc. He presents an analysis of the individual differences of management staff and ways in which this individuality can be optimized for program efficiency. Because of the need for careful communication, some individuals should spend at least 50% of their time in communication and coordination on a person-to-person basis. If the modeling program is geographically diffuse, a considerable amount of time is required for traveling.

As a number of participants in the program increases linearly, the complex of communication linkages which require personal attention tends to expand exponentially. Therefore, a large integrated program should have an individual specifically designated as modeling coordinator. His motivation results from the possibility of developing his professional reputation in this capacity more rapidly than could be accomplished by teaching in a small college. He must be communicative, personable, able to stand up and tell people when they are wrong, and have the intellect to explain why they are wrong.

Unless this modeling coordinator is isolated from traditional reward systems, he may be ineffectual. If the coordination task is merely ancillary to more important duties such as teaching and publishing, he may lack the proper motivation. In addition, the researchers may feel he is trying to steal their ideas to put them in his own papers. If a suitable person can be found, he can be a key player in the integration of the modeling project.

MODELER - EXPERIMENTALIST INTERACTION

Beyond establishing an effective relationship between the leadership and modeling personnel, management has an obligation to foster the integration between modelers and other research participants which leads to effective interdisciplinary endeavors. Management must break down communication barriers and develop an _esprit de corps_, a pride in the total program.

Because of the importance of program integration, it may be useful to document three simple indicators of project integrity which may already be well known to the experienced manager. First, project personnel should be complaining to management, but defending the program to outsiders. Complaints should be viewed optimistically since it indicates that participants are communicating and feel that management can and will redress grievances. Real dangers result from silence to you and carping to others. Second, one should be sensitive to the use of the pronouns, "us" and "them". An understanding of who "us" represents to various segments of the program permits diagnosis of where major communication barriers remain. Third, a rudimentary understanding of nonverbal communication in the form of body language and physical contact between individuals is also invaluable in diagnosing where communication links are firm or weak. All of these behavioral indicators should be closely monitored, particularly during meetings of project personnel.

In general, meetings are a key ingredient in program integration. Large, total-project meetings are an effective mechanism for achieving identity with the program. The meetings are most effective in isolated places and with a fair degree of informality. Do not over-structure the meeting; provide ample time for people to meet each other and learn to know each other as individuals. However, if people need to exchange substantive information or work on a scientific problem, small, specific-purpose meetings are more effective. These meetings should be informal and conducted in environments conducive to real interpersonal contact. Quantity and quality of specific output from such meetings seems to improve with groups up to five participants, but then deteriorates rapidly with increasing size.

Large-scale projects are ordinarily divided into functional subsystems for practical reasons such as evaluation or synthesis. Careful structuring of these subsystems can facilitate integration. The structures should be project- or goal-oriented, and not discipline-oriented. The latter fosters the "us-them" phenomenon. Modelers should not meet together; scientists dealing with a specific process such as decomposition should meet together, including modelers, microbiologists, botanists, soil scientists, etc.

Total program subdivisions can also be structured to develop mutual dependence. The simple mechanism of encouraging research on a specific aspect of the problem at the institution best qualified to do the research disperses the total effort and prevents each research group from competing with each other in the solution of the total problem. If each member of the team recognizes that he cannot adequately perform his task without depending upon the others, communication is fostered and cooperation becomes natural.

Problems between modelers and field researchers begin when the experimentalist is relegated to the position of technician. This is in obvious and direct conflict with his motivational structure. The system must not make him feel that he is merely supplying numbers for someone else, the modeler. The modeler will then have the opportunity to do the truly challenging intellectual task of developing higher level insights. From his own standpoint, the experimentalist would rather use the modeler simply to analyze his data. Then the researcher could publish the results with appreciative mention in the acknowledgments. This motivational conflict is intensified by physical distance since modelers may convey an image of dropping by to pick up hard-won data, then going away to do something mystical with it. The scientist's stock in trade is his insight and intuitive grasp of the system. He sells concepts and understanding. The problem develops when the modeler seems to be saying, "Tell me all you know about _____ in fifteen minutes. I want to extract your best insights and make my reputation with them." Does the egotistical experimentalist gain anything from such an interaction? A consideration of his motivational structure suggests not.

It is obvious from this discussion that effective communication between participants requires mutual trust and respect. It most frequently falls to the modeler to initiate the interaction and to develop the trust. Initiate by offering to help, by demonstrating the benefits of the interaction to the researcher, by relating to his motivations. This personal interaction skill is an important part of the modeling art, and may be as important in selecting modelers as their skill with mathematics or computers.

In our program, several approaches have been successful in establishing effective interaction. The "critical mass" approach utilizes a central pool of modeling talent available to the entire project. This group is used to concentrate on specific problem areas for limited periods of time and then move on. This core group is soon welcomed by the components of the program because they always come to give something, not to take something away. They leave models; they do not remove data. Also, they are always responding to requests for assistance rather than being forced upon the investigators.

The "impressive leader" approach utilizes the unique individual who is master of his game and who engenders cooperation because of the respect that people have for him. He relies upon an ability to stand face to face and argue from a viewpoint of knowledge and scientific intellect. Scientists will always respond more favorably to this approach than to budget manipulations.

The "team" approach relies upon extensive personal contact and communication between modelers and experimentalists. There is physical proximity, and the modeler comes to be considered as another member of a team of scientists. This approach is doubtless the most effective and practical, because if a real team does develop, the modeler-experimentalist interface disappears. There exists simply a group of scientists working together. This approach tends to break down the artificial distinction which raised the problem in the first place. The ultimate resolution is true integration in which the "model" is as much the product of one member of the team as another and data collection cannot be viewed as proprietary. When no one can remember for sure who originated a specific equation or who was responsible for a specific experimental design, the interface dissolves and integrated research becomes a reality.

CONCLUDING REMARKS

I have attempted an analysis of the problems associated with management of large-scale environmental modeling projects. A structure was provided for the analysis by considering the nature of modeling and the characteristics of the modeler as a scientist. Particular emphasis was placed upon the problems that arise when program objectives required a high degree of creative input from scientists that were situated across a wide geographic area.

As mentioned in the introduction, some aspects of this analysis will not be relevant to every modeling project. The structure of the total program, requirements of funding agencies, and overall program objectives will strongly influence all aspects of the project, including the modeling. Each program is likely to evolve slightly differently, and it is important to recognize several characteristics of the EDFB which affected the emphases in the analyses.

The EDFB became widely diffused geographically in order to involve as much of the available ecological talent as possible. This diffusion resulted in problems of communication, program integration, and conflicting reward systems. These problems, therefore, are understandably emphasized here. It should be obvious that a different spectrum of problems will arise when the modeling project is housed at a single institution. In that case, major problems may revolve around interactions between the modeling group and other participants in the program. But the internal management problems will appear quite different.

In order to initiate a massive program to understand forested ecosystems (still poorly understood), creativity played a critical role in addressing program objectives in the EDFB. This requirement for creativity led to an emphasis on the individualistic scientist working in the university. In other projects, objectives may stress a logical, technically competent application of results. Under these circumstances, the typically creative personality may be more a detriment than an asset.

So I must stress at the conclusion, as I did at the beginning, that the reader must determine the aspects of the present analysis which are relevant to his own program. Insofar as the characteristics of a specific program fit the pattern of the EDFB, more aspects of the analysis will seem applicable. It is hoped that some portions of the discussion will be useful.

Having passed through a morass of problems and dilemmas with only tentative solutions which may be workable in specific contexts, the reader may be left with a sense of pessimism. I would prefer to conclude on a more optimistic note. The pessimistic attitude assumes that all problems must arise and adopts a negativistic design to prevent the worst from happening.

In fact, the overall impression I have gained from my involvement in an integrated systems analysis program has been highly positive. It has been a most rewarding experience, intellectually stimulating and exciting throughout. If the basic characteristics of a large-scale computer model are clearly understood and the peculiarities of the motivational structure of the modelers are considered explicitly, the management structure can provide a stimulating environment conducive to creative science, while providing strong and unambiguous leadership.

The management challenge is to establish a motivational environment that fosters productive work rather than imposes constraints. The system should establish its reward structure to maximize the probability of success instead of minimizing the probability of failure. Because of the nature of the creative process, such a positive viewpoint is essential.

Acknowledgments

I would like to thank the many modelers who have been involved in the EDFB for their patience with me and apologize to them for the many management errors I have made. They were the guinea pigs during the evolution of these concepts. I would also like to express my appreciation to R. L. Burgess, S. I. Auerbach, D. D. Huff, J. M. Neuhold and others who patiently labored with me to make this paper intelligible.

Research supported in part by the Eastern Deciduous Forest Biome, US-IBP, funded by the National Science Foundation under Interagency Agreement, AG-199, 40-193-69 with the Atomic Energy Commission - Oak Ridge National Laboratory, and in part by the U.S. Atomic Energy Commission under contract with the Union Carbide Corporation.

LITERATURE CITED

Auerbach, S. I. 1969. Research design and analysis of the structure and function of ecosystems in the Deciduous Forest Biome. IBP biome research proposal to the Environmental Biology Program, National Science Foundation.

Bartell, S. M., T. F. Allen, and J. F. Koonce. 1973. Multivariate analysis of species replacement dynamics in lacustrine phytoplankton. Eastern Deciduous Forest Biome Memo Report 73-38. University of Wisconsin, Madison, WI. 57 pp.

Baumann, P. C., J. F. Kitchell, J. J. Magnuson, and T. B. Kayes. 1974. Lake Wingra 1837-1973; a case history of human impact. Trans. Wis. Acad. Sci., Arts, Lett. (in press)

Brewer, G. D. 1973. Politicians, bureaucrats and the consultant. Basic Books, New York. 292 pp.

Burgess, R. L. and R. V. O'Neill (editors). 1974. Eastern Deciduous Forest Biome. Progress Report and Continuation Proposal. EDFB-IBP 74-1, Oak Ridge National Laboratory, Oak Ridge, TN. 730 pp.

Chen, C. W., and G. T. Orlob. 1972. Ecological simulation for aquatic environments. Final report for Office of Water Resources Research, U.S. Dept. Interior. 156 pp.

Crawford, N. H., and R. K. Linsley. 1966. Digital simulation in hydrology: Stanford Watershed Model IV. Dept. Civil Engineering, Stanford Univ., Technical Report No. 39.

DeAngelis, D. L., R. A. Goldstein, and R. V. O'Neill. 1973. A model for trophic interaction. Eastern Deciduous Forest Biome Memo Report 73-82. Oak Ridge National Laboratory, Oak Ridge, TN. 40 pp.

DiToro, D. M., D. J. O'Connor, and R. V. Thomann. 1970. A dynamic model of phytoplankton populations in natural waters. Report of the Environmental Engineering and Science Program, Manhattan College, New York, NY. 63 pp.

Drucker, P. F. 1974. Management: tasks, responsibilities, practices. Harper and Row, New York.

Fulkerson, W., W. D. Shults, and R. I. Van Hook. 1974. Ecology and analysis of trace contaminants. ORNL-NSF-EATC-6, Oak Ridge National Laboratory, Oak Ridge, TN. 446 pp.

Goldstein, R. A. and J. B. Mankin. 1971. Space-time considerations in modeling the development of vegetation. pp. 87-97 in C. E. Murphy, J. D. Hesketh and B. R. Strain (eds.), Modeling the growth of trees. EDFB-IBP-72-11. Oak Ridge National Laboratory, Oak Ridge, Tenn. 199 pp.

Goldstein, R. A. and J. B. Mankin. 1972. PROSPER: a model of atmosphere-soil-plant water flow. pp. 1176-1181 in Proceedings, 1972 Summer Computer Simulation Conference ACM, IEEE, SHARE, Sci. San Diego, CA.

Goldstein, R. A. and W. F. Harris. 1973. SERENDIPITY: a watershed level simulation model of tree biomass dynamics. pp. 691-696 in Proceedings, 1973 Summer Computer Simulation Conference, AIChe, ISA, SHARE, Sci, AMS, Montreal, Quebec.

Holling, C. S. 1959a. Some characteristics of simple types of predation and parasitism. Can. Entomol. 91: 385-398.

Huff, D. D. 1968. Simulation of the hydrologic transport of radioactive aerosols. PhD Thesis, Stanford University, Stanford, CA.

Huff, D. D., D. G. Watts, O. L. Loucks, and M. Teraguchi. 1970. A study of nutrient transport with the Stanford Watershed Model. Eastern Deciduous Forest Biome Memo Report 70-1, Univ. Wisconsin, Madison, WI. 30 pp.

Huff, D. D. 1972. HTM program elements, control cards, input data cards. Eastern Deciduous Forest Biome Memo Report 72-13. Univ. Wisconsin, Madison, WI. 46 pp.

Huff, D. D., J. F. Koonce, W. R. Ivarson, P. R. Weiler, E. H. Dettmann, and R. F. Harris. 1973. Simulation of urban runoff, nutrient loading and biotic response of a shallow eutrophic lake. pp. 33-55 in E. J. Middlebrooks, D. H. Falkenborg and T. E. Maloney (eds.), Modeling the eutrophication process, Utah Water Resources Laboratory, Utah State Univ., Logan, UT. 228 pp.

Ivarson, W. R., J. E. Jacques, and D. D. Huff. 1972. Report on implementation of lake and reservoir flow routing into the HTM. Eastern Deciduous Forest Biome Memo Report 72-135, Univ. Wisconsin, Madison, WI. 31 pp.

Ivlev, V. S. 1961. Experimental ecology of the feeding of fishes. Yale Univ. Press, New Haven, CT.

Jacques, J. and D. D. Huff. 1972a. Open channel flow simulation with the hydrologic transport model. Eastern Deciduous Forest Biome Memo Report 72-134. University of Wisconsin, Madison, WI. 19 pp.

Jacques, J. E. and D. D. Huff. 1972b. Snow accumulation and melt simulation. Eastern Deciduous Forest Biome Memo Report 72-136. Univ. Wisconsin, Madison, WI. 17 pp.

Kitchell, J. F., J. F. Koonce, R. V. O'Neill, H. H. Shugart, J. J. Magnuson, and R. S. Booth. 1974. Model of fish biomass dynamics. Trans. Am. Fish. Soc. (in press)

Levins, R. 1966. The strategy of model building in population biology. Am. Sci. 54: 421-431.

Lotka, A. J. 1925. Elements of physical biology. Williams and Wilkins, Baltimore, MD.

Ludwig, S. 1974. Turning weaknesses into strengths. Sky Magazine, May 1974, pp. 21-23.

MacCormick, A. J. A., O. L. Loucks, J. F. Koonce, J. F. Kitchell, and P. R. Weiler. 1972. An ecosystem model for the pelagic zone of Lake Wingra. Eastern Deciduous Forest Biome Memo Report 72-122, Univ. Wisconsin, Madison, WI. 103 pp.

Mar, B. W. and W. T. Newell. 1973. Assessment of selected RANN environmental modelling efforts. Report prepared for RANN, Environmental Systems and Resource Division, National Science Foundation.

Minorsky, N. 1962. Nonlinear oscillations. D. Van Nostrand Company, Inc., Princeton, NJ.

Murphy, C. E., Jr. and K. R. Knoerr. 1970. A general model for the energy exchange and microclimate of plant communities, pp. 786-797 in Proceedings, 1970 Summer Computer Simulation Conference, ACM, IEEE, SHARE, Sci, Denver, CO.

O'Neill, R. V. 1974. Modeling in the Eastern Deciduous Forest Biome. In B. C. Patten (ed.), Systems analysis and simulation in ecology, Vol. III. Academic Press, New York. (in press)

Park, R. A., R. V. O'Neill, J. A. Bloomfield, R. A. Goldstein, J. B. Mankin, H. H. Shugart, R. S. Booth, J. F. Koonce, M. S. Adams, L. S. Clesceri, E. M. Colon, E. H. Dettmann, J. A. Hoopes, D. D. Huff, S. Katz, J. F. Kitchell, R. C. Kohberger, E. J. LaRow, D. C. McNaught, J. L. Peterson, D. Scavia, J. E. Titus, P. R. Weiler, J. W. Wilkinson, and C. S. Zahorcak. 1974. A generalized model for simulating lake ecosystems. Simulation. (in press)

Passioura, J. B. 1973. Sense and nonsense in crop simulation. J. Aust. Inst. Agric. Sci., September 1973, pp. 181-183.

Reeves, M. and E. E. Miller. 1974. Application of proportioning variant model of moisture hysteresis. Water Resources Res. (in press)

Shugart, H. H., R. A. Goldstein, R. V. O'Neill, and J. B. Mankin. 1974. TEEM: a terrestrial ecosystem energy model for forests. Oecologica Plantarum. (in press)

Smith, F. E. 1952. Experimental methods in population dynamics: a critique. Ecology 33: 441-450.

Sollins, P., N. T. Edwards, and W. F. Harris. 1974. Organic matter budget and model for a southern Appalachian Liriodendron forest. In B. C. Patten (ed.), Systems analysis and simulation in ecology. Vol. 4. Academic Press, New York. (in press)

Strauss, G. and L. R. Sayles. 1972. Personnel: the human problems of management. Prentice-Hall, Englewood Cliffs, NJ. 684 pp.

Van Dyne, G. M. 1972. Organization and management of an integrated ecological research program. pp. 111-172 in J. N. R. Jeffers (ed.), Mathematical models in ecology. Blackwell Scientific Publications, Oxford.

Volterra, V. 1928. Variations and fluctuations of the number of individuals in animal species living together. J. du Conseil intern pour l'explor de la mer III. Vol. 1. Reprinted in R. N. Chapman (ed.), Animal ecology. McGraw-Hill, New York, 1931.

Walsh, J. J. 1972. Implications of a systems approach to oceanography. Science 176: 969-975.

Water Resources Engineers, Inc. 1968. Prediction of thermal energy distribution in streams and reservoirs. Final report to California Department of Fish and Game.

Watt, K. E. F. 1959. A mathematical model for the effect of densities of attacked and attacking species on the number attacked. Can. Entomol. 92: 129-144.

Wiegert, R. G. 1973. A general ecological model and its use in simulating algal-fly energetics in a thermal spring community. pp. 85-102 in P. W. Geier, L. R. Clark, D. J. Anderson and H. A. Nix (eds.), Insects: studies in population management. Ecol. Soc. Aust. (memoirs 1), Canberra.

Present Problems and Future Prospects
of Ecological Modeling

Gerald T. Orlob

INTRODUCTION

The primary charge to the participants in this symposium was to
explore, in substantial depth, the role of ecologic modeling in a
resource management context. In his opening statement, Dr. Russell
emphasized the need to stimulate a "convergence" of ideas, not only
as exemplified by the symposium itself, but on the part of decision
makers in their stewardship of the environment.

Public intervention in the decision process, so evident in the
environmental movement of recent years, has forced tradition-bound
planners and designers to widen the range of alternative actions and/or
strategies to be assessed before the final choice is made. Not only
must we increase the number of viable alternatives that the body politic
desires to consider, but we must embrace more factors that bear on the
response of the environment to any specific alternative. Today, we
also see the environmental system with greater clarity than ever before,
with enhanced awareness of its complexity, its sensitivity and adaptability to exogenous forces, and of our own limitations in understanding
of how it actually behaves. This awareness is humbling to say the
least, but the very challenge presented has stimulated the more adven-
turous to find new tools, techniques or methodologies for dealing
quantitatively with resource management alternatives. The ecologic
model, the specific focus of our discussions in this symposium, is one
such "tool."

This concluding presentation in the symposium is intended, first, to serve as a review or recapitulation of the major questions and issues that have been developed by the speakers and the discussion stimulated by their presentations. Second, it will serve as a vehicle for individual commentary by the author, from his own experiences in the field of resource management, and specifically with regard to mathematical models of the aquatic environment. Finally, it is intended to identify areas of uncertainty or deficiency in the state of our ecologic modeling art and to point the way toward improving our ability to use the ecomodel in a meaningful way in resource management. We will conclude with some positive recommendations for future action.

THE MANAGEMENT CONTEXT

Implicit in "resource management," in the way we have considered it in our present discussions, is a search for optimality. We are deliberately seeking to find that solution or strategy for our stewardship of the natural environment, as influenced by man, that most nearly satisfies man's requirements. It is an exercise in optimization for the public good, however it may be defined.

Dr. Spofford, in his theme paper, focuses our consideration to those resources that are by their nature "common property," the air, the land, and the water, and which may be overused to the overall detriment of the general public. This focus, by its very broad nature, necessarily raises some knotty problems in the development of ecologic models to facilitate management decision making. For example, what are to be the objectives in the search for optimality? In the mathematical modeling world we customarily cast our objectives in mathematical terms, i.e., as objective functions. These may be explicitly embedded in the framework of the model itself, or they may be exogenous to the model, serving to guide the application of the model. In the former case examples are the use of economic criteria like least cost, maximum net economic benefit, or maximum benefit-cost ratio that may serve the requirements of linear or dynamic programming, or the like. In the latter instance, an example might be found in the application of the model for an organized evaluation of the response of the marine environment off New York Harbor to specific proposed alternatives for dumping of digested sludge from metropolitan New York. In this case, we may presuppose that the shear complexity of the environment precludes any simple solution and that the role of the model is one of elucidation of environmental consequences of any specific management strategy or alternative.

In acknowledging the role of models in the management process we must also recognize their inherent limitations as mere representations of the real world. The results we may derive from their application are often only a small quantitative part of the input to the decision process. Other equally important inputs may be purely qualitative, or may be inferences drawn from the model that are of themselves subjective. Moreover, since societal values change, so must the objectives or objective functions we may use, hence, the model must be viewed as a flexible tool, adaptable to each new set of circumstances as it is presented.

WHO ARE THE DECISION MAKERS?

The model maker is invariably the greatest enthusiast for his art
and its alleged potential to serve the decision process. Often he
becomes the decision maker in his own right and the model becomes an
end in itself. Such may be the case when the model serves only as an
educational tool. However, in the broader context we are seeking to
develop tools for resource management, to aid nonmodelers in the "big"
decisions, i.e., to build or not to build, to discharge or not to
discharge, etc. In this context we may regard the decision makers as
relatively uninformed on the structure of models *per se* but, nonetheless,
cognizant of the models' worth (or lack of same) as adjuncts to decision
making. The degree to which persons who may rely on models are made
aware of capabilities and limitations may well determine the viability
of the model as a meaningful component part of the resource management
package. This is a matter of education--of communication--between model
developer and model user.

Those who use models range from the academic, who tends to look
upon them as teaching and research aids, to the politico, who as guardian
of the public trust could care less how results are obtained as long as
they are politically expedient. In between these extremes, neither of
which are represented in pure form in this symposium, are arranged
those sincere exponents of the modeling art who actually see it as
serving in the achievement of practical objectives. These are divided
roughly into four groups:

1. The model developer-users who conceive the model, formulate
    and program it, and who see it carried through to at least

its first application to the prototype. This group is well
represented by the participants in this symposium.

2. The analyst-users who take the model package, adapt it to
their practical problems, and apply it more or less routinely
with the objective of finding an acceptable solution. Quite
a few of our participants are in this category.

3. The planner-managers who formulate problems, interpret model
output, digest them, and present summarized results to those
who make the decision. A very few of those at this symposium
are in this category although it may well be a role others of
us should aspire to secure.

4. The decision makers, who consider the reduced spectrum of
results presented to them and make choices. No one of those
participating in the symposium is in this category.

It is of value, I believe, to remind ourselves that we customarily
exclude the decision maker from our discussions, as in the present
instance. This occurs either by intent or inadvertently. Small wonder
that our product is so poorly appreciated by those we would like most
to impress! We surely need to do a better sales job; i.e., improve our
communication up the line to where the decisions are really made. This
may be the best thing we could do to advance the cause of new developments in modeling technique.

WHO ARE THE MODELERS?

We have implied that the modeler is not the decision maker; but
rather is one who is involved at the lower (beginning) end of the ladder

leading to decisions in resource management. He is often an academic by nature, since the model development process is of sufficient intellectual challenge to draw his interest. While the modeler may possess substantial mathematical skills, often to a very high level of competence in practical application, he is not merely a programmer. He invariably is a skilled professional or scientist in his own right, who sees in the modeling process the opportunity to formalize quantitatively certain basic physical principles, as well as empirical relationships, that characterize environmental behavior. He uses the model to translate ideas into actions, to convert data and information into dynamic descriptions of phenomena, and to organize complex interactive processes and behavioral mechanisms into a consistent functioning system. His goals in the modeling exercise may vary from the sheer intellectual delight of creativity to the pragmatism of putting the models to work as tools for decision making.

Originally, the modeler was an engineer, mathematician, physicist, or other scientist, with a strong quantitative orientation who confined his efforts to his own discipline. However, as the potentials of the model approach were broadened and extended to more complex systems, the activity became multidisciplinary. Specialists were needed to define the phenomenological relationships to be modeled, to put these into functional mathematical forms, and to provide essential empirical relationships that transformed the prototype into a mathematical model. Often these specialists were not modelers themselves, but were drawn into the circle of model builders to provide the credibility that only basic scientific grounding and experience can assure. They became members of a modeling "team."

A key member of the interdisciplinary modeling team is one who, in addition to providing certain essential skills, has a broader view of the modeling domain. He must be able to synthesize, conceptualize, and integrate the component parts into a "realistic" representation of the prototype. Often this is a person of broad experience, perhaps not directly involved with the mechanics of computer applications, but who has the appreciation of the model's potentials for meaningful use in the management context with which this symposium is concerned. He is the leader of the team effort. He falls most often into categories 2 or 3 of the decision hierarchy outlined previously.

Aquatic ecologic modeling is surely one of the most demanding of all from the multidisciplinary point of view. Consider, for example, the general conceptual layout for an ecologic model of the New York Bight, as illustrated in Figure 1. By implication in this schematic representation, the scientific disciplines and technological skills required would include at least those of physical, chemical, and biological oceanography, meteorology, aquatic ecology, fisheries, marine geology, hydrology, hydrodynamics and environmental engineering. Considering, also, that the ultimate objective may be to develop the conceptual model to a level where it can be used in a management context, it would be desirable to have the participation of a resource economist, a water resource planner, and certain specialists in the computer sciences. Truly, ecologic modeling is not a one-man job! The experiences recounted in this symposium of the RPI team in developing CLEAN, of Hydro Science in its modeling work on the Great Lakes and various estuarial systems, and of WRE in developing its lake and estuary ecologic models, strongly support this observation.

290

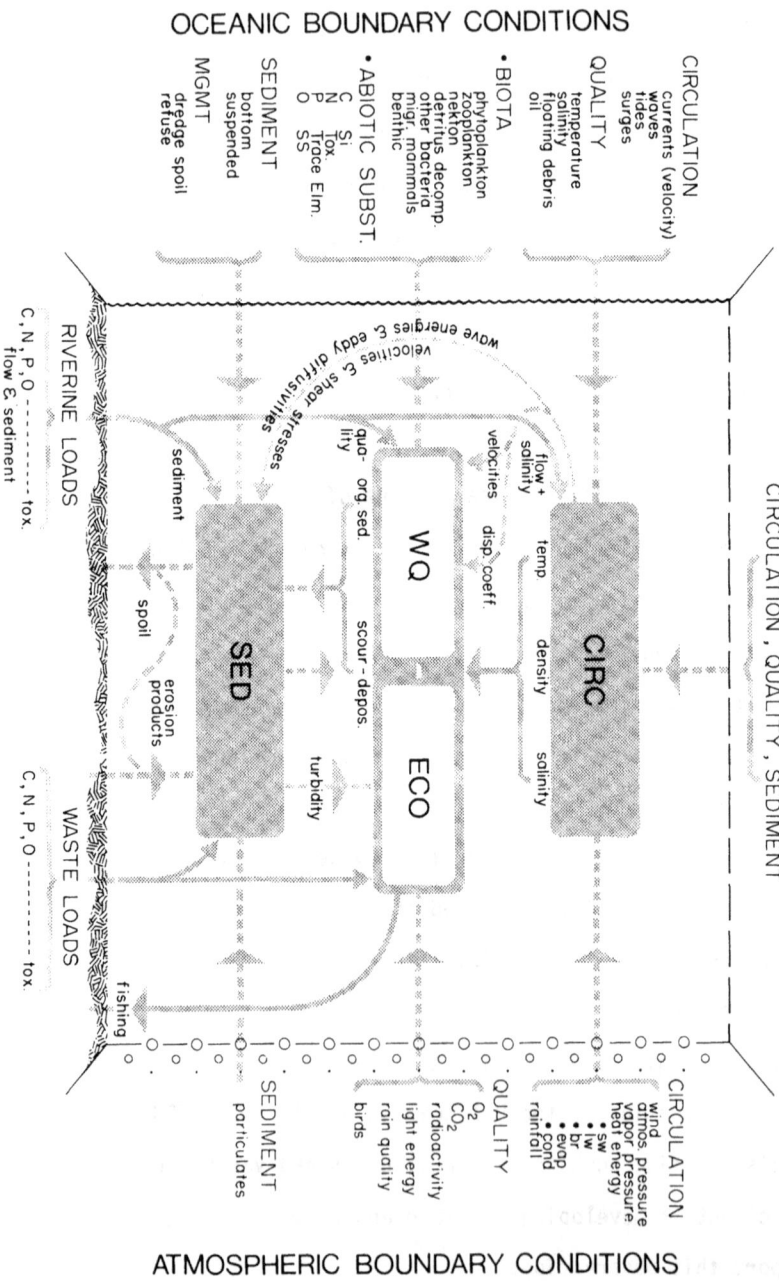

Figure 1. GENERAL CONCEPTUAL MODEL OF A MARINE ECOSYSTEM

MODEL DEVELOPMENT PROCESS

Before reviewing the specific problems of ecologic development that have been cited by the previous authors and discussants, it is appropriate to outline briefly the process of model development. With this in mind we may be able to be a bit more explicit about just where we stand at the present.

The technique of mathematical model development is in reality an art, and the model tends to be a product of the creativity of the developer(s). It is a characterization of the prototype, among many that may be conceived, and in this sense it is virtually always unique, seldom general. Nevertheless, there is a certain order and discipline in the process of model development, just as there may be in other artistic endeavors. This process, as it is customarily practiced, is schematically represented in Figure 2 and described briefly below.

The process begins with a definition of goals or objectives to be reached, usually what the model is to be used _for_ rather than what explicit form it will take. For example, this statement of intent may stipulate that the model is to guide the process of long range planning by representing the year by year average response of the proto-type to stimuli resulting from population growth, e.g., eutrophication as a result of increased nutrient loading. The time and space scales for simulation can be prescribed at this point, although usually they are only broadly indicated; the developer is left to decide. Further, it may be stipulated that the response of the prototype as predicted by the model is to be measured in terms of certain parameters (algae

GOALS & OBJECTIVES

prototype
descriptions → CONCEPTUALIZATION

space and
time scales → FUNCTIONAL
REPRESENTATION

solution
techniques,
hardware → COMPUTATIONAL
REPRESENTATION

functional model

test case data
coefficients
boundary conditions → CALIBRATION

calibrated model

MODEL
SENSITIVITY

verification data → VERIFICATION

verified model

test case
results → DOCUMENTATION → USER
MANUAL

documented model

feedback to
improve model

prototype
data → APPLICATION

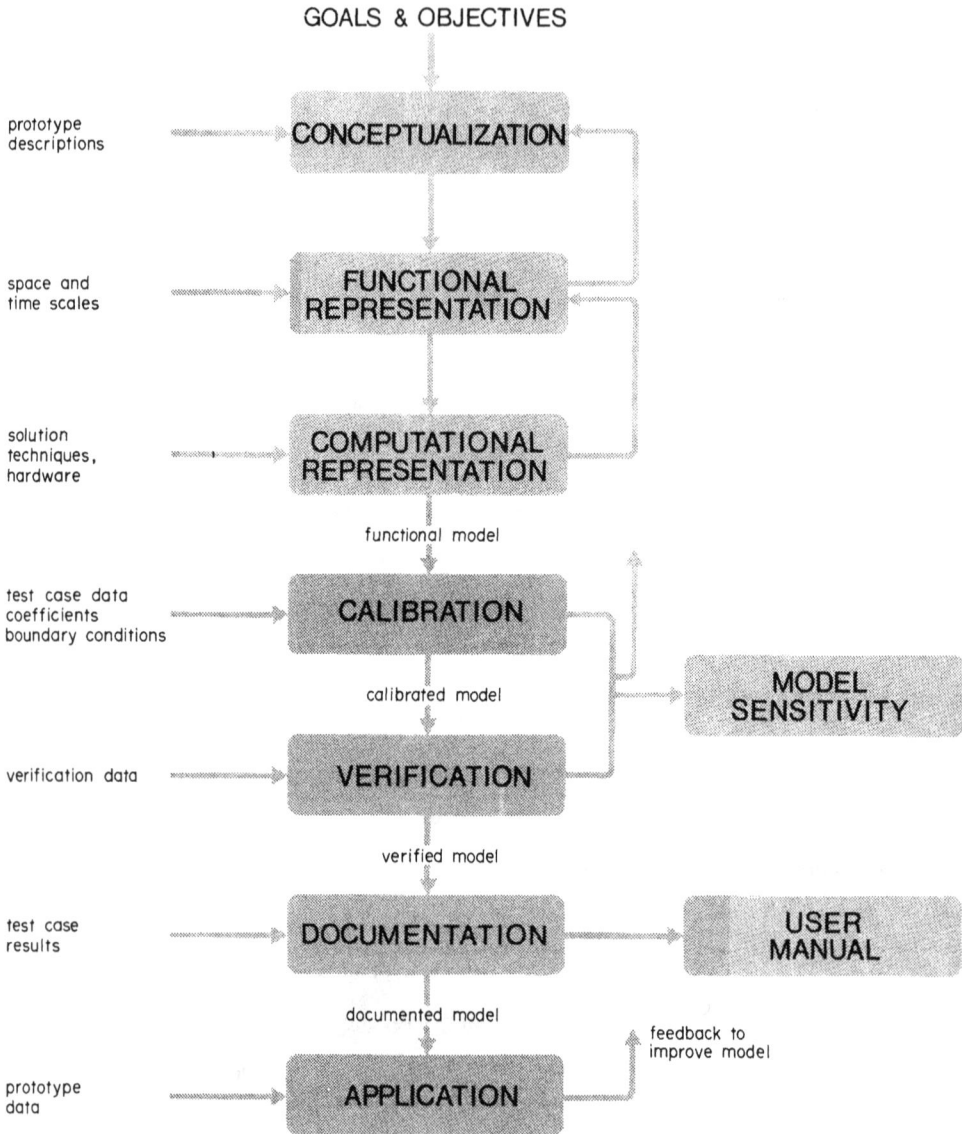

Figure 2. MODEL PROCESS DEVELOPMENT

biomass, chlorophyll, dissolved oxygen, total nitrogen, sulfite waste
liquor, etc.) that are meaningful to the ultimate user of the model.

Once goals and objectives are defined, the model development process
proceeds through a sequence of phases, more or less logically ordered.
For a specified model development--that is, a development program
spelled out under contract--there are six major phases or steps:

> Conceptualization,
>
> Functional Representation,
>
> Computational Representation,
>
> Calibration,
>
> Verification (Sensitivity), and
>
> Documentation.

Once the specific model package is placed in use, i.e., applied,
it may be further modified, corrected or improved. This phase in the
process is actually outside the model development process, *per se;* it
is an ongoing activity wherein the user changes his goals and strategies
and adjusts his model needs accordingly.

In order to place the model development process in perspective,
let us describe briefly each of the phases involved.

CONCEPTUALIZATION

Conceptualization entails primarily the physical representation of
the prototype water body in three spatial dimensions. Consideration is
given also to rates of change and the magnitudes of variation in spatial

dimensions, e.g., changes in water depth, hence time scales are important. Usually this phase of the process involves a discretization of the water body into segments or elements of such size and dimension that in the aggregate assembled as a "system," they take on the appearance of the prototype. Generally, volume and mass continuity are preserved. The technique of representing an estuarial system by "nodes" (volume elements) and links (transfer elements) is an example. Information required in this phase consists usually of hydrographic maps and bathymetry for the water body being modeled.

FUNCTIONAL REPRESENTATION

Functional representation entails mathematical formulation of the equations that govern the phenomenological behavior of the prototype. Usually, these will be in differential form applicable to each element of the system and in the dimensions required for suitable representation. Information on time and space variations expected in the prototype are needed at this point in order to determine whether gradients are of such small magnitude that some equations may be eliminated or, if not, at least simplified. For example, intense vertical mixing in a tidal estuary may justify the elimination of the equations of mass transport along the vertical axis, a gradually varied flow in a stream system may justify an approximation of steady state, or the minor fluctuation of temperature throughout a daily cycle may justify a consideration of biological reaction rates independent of temperature. The product of this phase of model development is usually a set of differential equations in space and time.

COMPUTATIONAL REPRESENTATION

Since formal, closed solutions of the set of equations may not be possible or practical in view of the number of elements and the time dependent factors being considered, the equations are usually converted to numerical form for solution with the aid of a computer. This phase entails reformulation of the equations into numerical computational form and programming for solution. Consideration is given in this phase to the specific numerical techniques and the computer hardware to be used. These may, in turn, require reconsideration and adjustment in the time and space scales, and in the number of dimensions considered in the model. Actually, there is a substantial "feed back" within the first three phases that results in creating a model that will best meet the objectives within the limits of technical and computational capability. The skill of the modeler is the determining factor in securing the best model for the job.

The product that emerges from this phase is a so-called "functional model" capable of prototype simulation, given reasonable values of coefficients, initial conditions, and boundary values. However, it is not yet checked against the prototype; it merely functions in a rational fashion as subjective experience with the prototype suggests.

CALIBRATION

In order to truly qualify as a model, the functional model must be adjusted to the prototype with "hard" data. A test case is selected that best exemplifies prototype behavior and for which real observational data are available. Since the model will invariably include some empiricism, in the form of coefficients, numerical approximations, or assumed boundary conditions, it will be necessary to "tune" it to perform like the prototype. This is usually accomplished through a succession of trials until the modeler is satisfied with performance. Criteria for calibration are often rather subjective, although some attempts have been made recently to make this process more rigorous (1). The product of this phase is a calibrated model capable of representing prototype behavior for a single set of given data.

VERIFICATION

The real test of the model's capability lies in its ability to simulate prototype response for a new set of given conditions, i.e., new data, substantially different from those used in calibration. The model is subjected to this test and results are compared against the observed prototype response, using the same acceptance criteria that were employed in the calibration phase. If the criteria are met, the model is considered "verified," or validated, and in condition to be used as a tool for planning. Here again it may be observed that the verification process tends to be largely subjective; no formal procedures have yet been widely adopted by modelers.

SENSITIVITY TESTING

As a part of the exercises of calibration and verification, that in themselves are closely related, the modeler may develop the bases for defining the "sensitivity" of the model's response to variations in input data, coefficients, boundary conditions and even the functional representations included in the model. The product of this effort may be a modified or improved model, or merely a statement of the model's limitations in terms of data and information supplied to it. Usually sensitivity would be expressed in absolute or percentage changes in a response parameter, e.g., dissolved oxygen, temperature, algal biomass, etc., to changes in an independent variable, coefficient, or exogenous input.

DOCUMENTATION

Once the model has been calibrated, verified and tested for sensitivity, it may be considered to be in condition for documentation. This step is a "must" in the process; it finalizes the model in the unique form in which it will be delivered to other users. Documentation consists of a complete description of the computer program, the subroutines, the computational logic, all pertinent variables and coefficients, and card formats, i.e., everything needed by a future user, unfamiliar with the model's development, to comprehend completely the genesis of the model and its capabilities and limitations. At this point, a User's Manual may be produced to guide future users in implementing the program on the computer. It includes a complete program listing, card formats, simple instructions on program usage and a demonstration problem fully executed and illustrated. The manual would be, in essence, an explicit step by step procedure sufficient to ensure operation without the complete background, rationale, and logic that would be found in documentation.

The final product of the model development process is a fully documented model, tested and presumed ready for application.

APPLICATION

Actually, the model development process continues beyond documentation. In applying the model to real problems, the user may discover fallibilities or limitations not known or anticipated by the developer. These may range from minor errors that can be corrected to conceptual errors that may call for major program revisions. In the latter case, one may have to repeat all or several of the steps outlined above. Inasmuch as a "new" model could result, it would at least be necessary to document it fully and perhaps to recalibrate and reverify.

Experience indicates that application of the model is essential to define fully its capabilities and limitations. Despite the best intentions of the developer, the model is always fallible; after all, it is only a representation of the prototype, not the prototype itself.

PRESENT STATUS IN MODEL DEVELOPMENT

Where do we stand at the present in this process of development of ecologic models? More specifically, do we now have the capability to apply ecologic models in a resource management context?

One gains the impression from talking to eco-modelers--especially those who have participated in this symposium--that the state-of-the-art is rather well advanced. Perhaps it has even reached the level of meaningful input to the decision making process, as the examples cited by Drs. Chen, O'Connor, Thomann, and the author of this paper seem to indicate. Nevertheless, all are anxious to acknowledge that much remains to be accomplished. Let us examine a few of the more obvious areas of deficiency in our present capability to model the aquatic ecosystem.

Data

Almost everyone agrees that we need more data. (In fact, this seems to be the case whether we are dealing with the question of models or not!) Invariably, there is a deficiency in data on proto-type response, i.e., in the data needed to calibrate the functional model and subsequently to verify it.

If one examines the body of information collected over the years on prototype systems, it is clear that the data needed to calibrate a model like CLEAN, for example, rarely exists. It is necessary for such purposes to initiate specially designed data collection programs, aimed at the particular preconceived requirements of the model. For example, data on the basic nutrients, nitrogen and phosphorus, and the standing

crops of algae and zooplankton, are seldom available in the amount,
form, and at the frequency necessary to corroborate a model simulation
of the prototype. Moreover, when such data are found, they are usually
not supported by corresponding observations of other important parameters
of the ecomodel, such as those related to water quality, hydrology, or
hydrodynamic behavior.

Aside from mounting comprehensive, and costly, collection programs
to fill in all the gaps in our data base for the most elegant of
ecologic models, it may behoove us to consider some compromises in
coping with this weakness in the state-of-the-art. Two approaches
seem attractive.

The first of these is exemplified by the work of Dr. O'Connor and
his associates who tend to adapt their models to the existing data
base, whatever it may be, rather than the reverse. This pragmatic
approach is predicated on the reality of the world of decision making
with limited budgets, short time frames, and explicitly defined work
programs. For obvious reasons, the model tends to·be simplified
consistent with the data base, and the level of detail required by
the decision makers. There is a tendency for aggregation, i.e., temporal
and spatial averaging.

An alternative approach, that could be developed with the aid of
the WRE Lake Ecologic Model, CLEAN or other so-called "comprehensive"
models of the aquatic ecosystem, would be to use the model as an aid
to design of the data collection program. Sensitivity testing, for
example, can be used to indicate the relative importance of improving
the data base for specific parameters or to determine the consequences of

eliminating entirely the need for certain data by modifying (simplifying) the model. A potentially attractive technique is one proposed
by Moore (2) in which a model is coupled with a statistical filter.
The combination can be used to improve the predictive reliability of
the package beyond that of the deterministic model by itself, producing
a quantitative estimate of reliability, reflecting both the statistics
of the data input and the structure of the model. The inverse problem
can also be solved, i.e., one can determine the frequency of sampling
or density of data needed to assure a predetermined predictive reliability.
To date, Moore's work has been confined to quite simple ecomodels and
linear filters, but the results seem sufficiently promising to justify
additional research.

Hydrodynamics, Hydrology and the Ecosystem

The circulation of water in the aquatic environment, governed
by the forces of nature and sometimes by man, is an important determinant
in the response of the ecosystem. In fact, evaluation of the ecologic
consequences of changes in circulation is often the central issue in
resource management. Examples cited in this symposium of the Sacramento-
San Joaquin Delta, the Great Lakes, and various man-made impoundments
highlight the importance of correctly representing the movement of water
within the aquatic habitat.

The degree of detail provided in hydrodynamic characterization depends, of course, on the specific objectives of the modeling effort. Steady state conditions may be reasonable for a river system above the influence of tide water, but are hardly appropriate in the estuary where flows are oscillatory. When variations in density occur, due to changes in temperature, salinity or suspended solids, the flows that transport nutrients and non-motile biota through the system are more difficult to define and the requirement for faithful representation of circulation becomes more stringent. The example of the stratified reservoir, the temperature profile of which may be strongly influenced by man's operation of the facility, is a case in point. Another is exemplified by the circulation of water in shallow embayments under the action of tide, wind, and salinity differences. It is the author's contention that adequate characterization of circulation is basic to the ecomodeling process; without it, the model cannot be made to serve the goals of resource management.

Ecological Concepts

There seems to be a consensus among the symposium participants that the principle of mass conservation and the empirically substantiated kinetics of chemistry and biochemistry are sufficient for conceptual and functional characterization of the aquatic ecosystem. The comparatively simple notions of mass continuity through a set of balance equations, coupled by first order interactions, seem to be attractive despite some dependence on empiricism.

This confidence in the approach seems to stem from the fact that there is a certain universality in the magnitudes of rate coefficients that have been derived by many independent investigators the world over. Moreover, the initial applications of ecomodels to prototype systems, using these coefficients, even though they were derived under more controlled environmental conditions, are especially encouraging. For example, the first application of the Lake Ecologic Model to Lake Washington, using rate coefficients selected from the literature, resulted in an almost remarkable agreement with the field observations made by W. T. Edmondson of the primary nutrients and algal biomass (chlorophyll a).(3) Similar experiences have been reported by Drs. di Toro, O'Connor, and Thomann in their applications of the Hydroscience ecomodel under a wide variety of environmental conditions.

There seems also to be a consensus that the conceptual design of the ecomodel should embrace, at the outset, the full complexity of the ecosystem, i.e., the entire food web. If there are simplifications to be made, for whatever pragmatic reasons, these should be made with full appreciation of the consequences to purity of biological representation. This is a concession, perhaps, to the biologists who are reluctant to accept the notions of aggregation and quantification that are so essential to the creation of a working mathematical model.

A troublesome area that has not yet been properly addressed, it seems to the author, is that of the so-called "perturbed," or collapsing ecosystem. Most modelers seem to have accepted the idea of

a dynamic equilibrium in the ecosystem, with perpetual repetition
of the life cycles embedded in the model. What happens if environmental
conditions deteriorate progressively to the point of failure of one
"compartment" of the system? If this occurs, there must be a major
readjustment in the flow of energy through the system, in fact, a
whole new ecosystem. The idea of "domains of attraction," advanced
by Holling (4) and the implication of the deliberate demise of an
ecosystem as a management alternative seem to be germane to this question. A challenging area of future inquiry for ecomodelers is strongly
suggested.

Aquaculture

One of the attractive potentials for the ecologic model in the
management context is the broadening field of aquaculture. The review
of the historic development of fisheries models presented by Dr. Schaaf
highlights this potential, in certain ways yet to be realized in
fisheries management, *per se*.

Not unlike single purpose planning and management efforts in other
quarters, the management of fish stocks has been oriented to benefit man
in the narrower sense. That is to say, that the primary objectives
have traditionally been to maximize catch, or harvest, potentials to
meet the commercial market demands. The primary focus has been on a
single trophic level, usually with relatively little consideration of
the remainder of the ecosystem. If other environmental factors were

considered by fisheries managers, they were usually taken as restraints
on the problem of optimizing, i.e., maximizing, fisheries production.

Today, it seems, the stage is set for a broader view of aquaculture,
both as regards the nature of the aquatic ecosystem itself and as
regards the benefits derived from management of the whole resource.
It has become increasingly evident from retrospective examination of
the fisheries industry's performance that failure to have considered
the aquatic ecosystem in the broad, more complete sense, accelerated
its demise in some quarters. Evidence appears now to indicate, for
example, that overfishing of the California sardine may have pushed
the ecologic balance beyond the "domain of attraction," whereas, a
reduction in fishing pressure at a certain critical stage might have
saved the industry.

Also, an increasing awareness of environmental factors, the small
"e" in Dr. Schaaf's differential equation, and their influence in
sustaining the viability of the food web is sure to lead to a broadening of the scope of fisheries models. The trend is surely in the
direction of more comprehensive ecosystem models as tools for management of the total fishery resource, not merely singular components
of the aquatic ecosystem.

A final encouraging note was added to the discussion of fisheries
models by Dr. Lackey, who pointed out that the perspective of "benefits
to man" is broadening also. No longer is it merely a question of the
quantity of fish harvested, but that other factors of a less tangible,
but nonetheless significant, sort now figure prominently in the
management of the total resource. These include other trophic levels

than the "target" as well as parameters pertaining to other beneficial water uses, among them mere esthetic enjoyment of the environment.

Acute vs Chronic Effects

A major, and perhaps justifiable, criticism of the current state of the ecologic art was sounded by Dr. Harris, that is, an overemphasis on near term response of the ecosystem as opposed to longer term response. The primary stimulus for the modeling effort, and related activities, has been to solve acute pollution problems, especially where there is a strong economic incentive to do so, at the expense of chronic problems slow to develop, for which there is no immediate payoff. Dr. Harris contends that a complete reorientation in our value system is required, away from economic objectives and toward the environment. The impression is given that since this is not currently the case, there is no value whatsoever in modeling efforts that are predicated on solving short term problems.

The author disagrees strongly with this notion for several reasons. First, there is no ingrained restriction in the resource management concept outlined by Dr. Spofford that precludes using any particular value system, economic or otherwise. The question is one of devising the objective function--or scoring system, or whatever--that will be used to rank alternative strategies. Second, there is nothing in the nature of the ecologic modeling approach that precludes consideration of chronic effects, when these can be properly represented functionally. It remains for the critics who consider this mandatory to assist the innovative modelers in taking this creative step forward. The author submits that positive action should supplant a negativistic, "do nothing" stance.

Prediction, Evaluation and Reliability

Virtually all of the ecologic models discussed in the symposium are descriptive simulation models. They require explicit sets of input data, predetermined boundary conditions, and estimated coefficients. In short, they solve a specific problem once stated; a new problem requires a new statement.

This characteristic, although it has certain advantages in giving insight into system behavior, also has its limitations when these results are translated for use in the real world of decision making. One limitation lies in the reliability of the model as a predictive tool. Since it is partly empirical, it should not normally be used to predict responses in the range very far beyond the limits of antecedent experience with the prototype. However, the author contends that, if the model is properly calibrated, and has undergone the test of verification as outlined earlier in this paper, it can be used to evaluate alternative consequences, probably with greater reliability than as an absolute predictor of future events. That is to say, the differences one notes between the model's predictions for alternatives A and B are likely to be of greater reliability than the absolute values of either. Thus, the model may suggest that B is a superior choice to A in certain ways, or vice versa. In this way it can be made to serve the decision process, giving information rather than making decisions itself. The decision maker will still have to choose, but hopefully now with more information at hand.

The second difficulty stems from our lack of attention to date
to matters of a stochastic nature. The models are now basically
deterministic and can only deal with risk analysis through conventional
simulation of selected sets of conditions. Because the models are
environmental by nature, involving many variables, some of them sure
to be interrelated, it is virtually impossible to formulate a representative set of test cases that will yield a probability distribution of
results that can be said to be that of the prototype. Some new techniques for statistical characterization of simulated response of the
ecosystem in relation to that of the prototype need to be developed.

Finally, we return to the question of collapse of the ecosystem
under the combined stress of multiple environmental factors, with
attendant alteration of the system itself. What set of critical conditions can cause this to occur and what is the probability that this
may occur? These questions are at the heart of the management problem;
answers are needed to set the bounds (restraints) for resource management, i.e., the limits for decisions.

Adaptability and Transferability

Considerable discussion ensued in the symposium session concerning
the transferability of a specific model to a site other than that for
which it was originally developed.

The hope of some modelers is that their product will be of universal
utility, merely requiring minor adaptation to a new site. In general,
this seems not to be the case; models are of a particular prototype,
not all prototypes, and until fitted with "site specific" data, the

ecologic simulator is not really a model at all. This is especially apparent when one considers the nature of physical, hydrologic, meteorologic, and other environmental boundary conditions that affect the ecosystem. Also, the process of calibration makes the model site specific; it should only be applied to that site from which the data for calibration were derived.

It is acknowledged, however, that there are some common features of ecomodels that are general and transferable, but these are imbedded in the concept of the model itself. The laws governing the interrelationship between trophic levels are universal in a sense, especially at the lower end of the trophic scale. It seems that bacteria and algae behave more or less the same the world over, and their growth rates under the same environmental conditions are generally reproducible within fairly narrow limits. This observation is substantiated by experience to date, although certain exceptions may later come to light. It is essential to recognize, however, that models may not always include that set of environmental conditions that necessarily governs the rates of growth, respiration, and mortality under all conceivable conditions.

Even the conceptual structure of the model may affect its applicability to other than a limited number of sites. For example, failure to include a particular predator, an alternative food source, or a critical environmental constraint may invalidate the model for a new situation where these may assume greater significance.

In short, the model must still be treated as site specific until it can stand the separate test of reverification for the new location.

Bacteria

Dr. Kelly and Dr. Nihoul both highlighted the need to give greater attention to the role of bacteria in the ecosystem. Special difficulties are created by the lack of specific experience with application of ecomodels that explicitly include bacteria. The experience cited for the Southern Bight by the Math Modelsea Group (5) seems to indicate that for some systems, at least, this is essential. An area for additional research is clearly indicated.

CONCLUSIONS AND RECOMMENDATIONS

The symposium provided a unique and timely opportunity for a few practitioners in the field of ecologic modeling to examine the state of their art in relation to its role in resource management. The discussion, always vigorous and candid, highlighted both the capabilities and the limitations of ecologic models as tools for decision making. It uncovered, as well, some promising areas for future research and development.

To conclude the symposium presentation on a positive note it is fitting, the author believes, to venture some specific recommendations, as follows:

1. Existing ecologic models should be calibrated, verified and applied to real situations to gain experience and understanding of the models' capabilities and limitations.

2. A formalized methodology should be developed for calibration, verification and testing of ecologic models to ascertain their reliability as predictive tools.

3. Field and laboratory experiments should be conducted under controlled conditions to improve estimates of rate coefficients in the governing mass transfer equations.

4. Controlled experiments should be performed in conjunction with models for equilibrium and perturbed (collapsing or expanding) systems.

5. Techniques should be developed for determining optimal combinations of models and the supporting data base, i.e., to design data programs to assure a prescribed reliability of a predictive model.

6. The anaerobic cycle should be included in existing models, i.e., a comprehensive submodel of the benthos should be developed.

7. The description of the bacterial compartment in ecologic models should be improved and supported by prototype data.

8. Existing fisheries models should be extended to the level of comprehensive ecologic (aquaculture) models, including food web and environmental factors.

9. Methods should be developed for considering statistically extreme ecosystem responses and assessment of risk of collapse in ecosystems under acute environmental stresses.

312

10. The effects of sublethal (chronic) stresses on the
     aquatic ecosystem should be incorporated into ecologic
     models and supported by reliable prototype data.

REFERENCES

(1) McLaughlin, Dennis, and Water Resources Engineers, Investigation
     of Alternative Procedures for Estimating Ground Water Basin
     Parameters, Office of Water Resources Research, Dept. of the
     Interior, to be published, December 1974

(2) Moore, S.F., "Estimation Theory Applications to Design of Water
     Quality Monitoring Systems," ASCE, Jour. Hyd. Div., V. 99, No. HY 5,
     pp. 815-831, 1973

(3) Chen, C.W. and G. T. Orlob, Ecologic Simulation for Aquatic Environ-
     ments, Office of Water Resources Research, U.S. Department of the
     Interior, Final Report, Project C-2044, December 1972, 156 p.

(4) Holling, C.S., "Resilience and Stability of Ecological Systems,"
     Annual Review of Ecology and Systematics, Vol. 4, pp. 1-23, 1973

(5) Nihoul, Jacques C.G., Math. Modelsea, Mathematical Models of
     Continental Seas, Dynamic Processes in the Southern Bight,
     International Council for the Exploration of the Sea, C.M. 1974 -
     c:1, 454 p.

EDITED DISCUSSION TRANSCRIPT

DAY ONE - MORNING

SPOFFORD PAPER: "An Introduction to the Framework"
(Interruption related to mention of Hollings' ideas.)

DR. ORLOB: These issues are really associated with the dynamics of
the system. For example, if the dissolved oxygen drops to zero--even for
a short period of time--and you have not constructed the environmental
model in such a way so that you can account for that event; for example,
if you have considered only a steady state, then you have missed the
event that causes the largest penalty.

DR. SPOFFORD: I agree. That is why I suggested that we try to incorporate environmental variations over time in our optimization, or
management, models. Peak concentrations may be at least as important
as the steady state levels. And, I think, eventually we will have to
deal with stochastic aspects--shock loads and weather--as well.

DR. CHEN: I would like to make a specific comment about the so-called
environmental zoo Dr. Spofford just mentioned.

In a recent study conducted for the Texas coast, we found that a
hurricane comes every so often and destroys the undesirable species. The
desirable species are then renewed by natural forces. Without human
management or some such natural cleansing action, it appears that an ecosystem will decay in such a fashion that the undesirable species will
eventually take over the entire system.

DR. SPOFFORD: That observation suggests that the dissipation of
hurricanes will have some interesting social and economic implications.

DR. CHEN: Yes. Our analysis indicated that in order to maintain
a good crop of certain organisms, hurricanes are required.

(Interruption related to discussion of explicit and implicit
forms of models of ecological systems.)

DR. KNEESE: Somehow, the explicit form of the environmental model
seems quite natural to me, and I can see how it comes about. But how
does the implicit form come about?

DR. SPOFFORD: When the partial differential equation set that is
characteristic of all mass-continuity environmental models is solved
using finite difference techniques, the following algebraic expression,
in matrix notation, obtains:

$$\frac{\Delta R}{\Delta t} = - f(R) + X \tag{1}$$

where R is a vector of state (endogenous) variables, X is a vector of
residuals discharges, and t represents the time dimension. For linear
environmental systems, Eq (1) may be rewritten, in matrix notation, as

$$\frac{\Delta R}{\Delta t} = - A \cdot R + X \tag{2}$$

For steady state conditions, the equation set may be reduced even
further.

$$0 = - A \cdot R + X$$

or rearranging terms,

$$X = A \cdot R \tag{3}$$

So that the implicit form results naturally enough. What we
actually want, though, is an explicit expression for the system state
variables, R. In order to obtain this, we must invert the matrix of
coefficients, A, premultiply both sides of Eq (3) by this inverted
matrix, and rearrange terms. That is

$$R = A^{-1} \cdot X \tag{4}$$

Note that A is a square matrix because one differential equation is
required to describe the time rate of change of each endogenous variable
considered in the model.

(General discussion following Spofford's paper.)

DR. ORLOB: I would like to make an observation and a philosophic statement. It has to do with the framework in which we are defining this discussion.

One of the characteristics of modelers is that they feel very comfortable once they have defined the boundary conditions of the problem. Then, they can crawl inside of the boundary and, of course, work very neatly on finding some solution to the management problem thus defined.

We do, however, have a lot of trouble defining those boundary conditions. In particular, we have difficulties in defining the economic boundary within which we are going to do this ecologic modeling. For example, who decides what the penalties are and who determines the form and attributes of the objective function that we use in the optimization model? Aren't these things just as crucial to the result that we get as the intricacy of the ecologic modeling itself?

I am not sure I know enough about economics to judge whether or not a given objective function is a socially desirable one. We might define any number of objective functions, and it seems that there is a high degree of subjectivity in imposing a particular criterion on the problem at the beginning. So, we have to be sure that we understand the framework and the boundaries within which we are working in order to be able to evaluate the ecosystem under varying conditions.

DR. SPOFFORD: This is an important point and something that I did not spend much time on in my presentation. It is an area that some of us at Resources for the Future, especially Ed Haefele, Allen Kneese, and Cliff Russell, have done quite a bit of thinking about and some research on.

First, let me explain something about the use of penalty functions that is critical to this discussion. Then I will ask Cliff Russell to address the issues of an appropriate objective function and the social choice mechanism. The penalty function is merely a mathematical device used in the non-linear programming algorithm to force the ambient environmental quality standards (constraints) to be met. The value of the penalty per se is not a social choice matter. The real issue facing society is how the ambient standards are selected. What process do, or should, we use to trade off environmental degradation in one place for a different type of environmental degradation in another place, or for different levels of material wealth? What is the distribution of costs, benefits, and environmental quality among geographic regions and among income groups for the various scenarios?

DR. RUSSELL: I think the social choice question is a good one to raise at the beginning. I suppose we are trying to trick you, in a sense, into talking about ecological models in isolation. If I can convey what I think the overall model is doing, it may provide an answer without me saying anything about the objective function.

I envision these management models as informing some kind of choice process. One can imagine estimating functions that relate dollars to some measure of "success" in a commercial fishery where "success" is in turn related to pollution levels. There have been attempts, which so far have been unsuccessful, to obtain similar functions for recreational fisheries, swimming, and general water-based recreation. If we had a world in which all the functions of this sort were known and people believed them, then the management models could be used to maximize a net benefit function which would mean something socially.

But first, I think we would all agree that the connections between ambient environmental quality and any economic measure is very tenuous at best. The aesthetic damages simply cannot be estimated now--and probably never can be. If we have estimates in a specific case, then they generally are not believed--and probably for good reasons--because they are frequently exercises in economic trickery.

From this point of view, then, the most meaningful criterion for ranking outcomes is cost, simply because we can measure the costs of treatment plants, sewer lines, changes in production processes and so on; while we cannot measure damage reduction. But in order to use cost, we must standardize the environmental output of the model. That is, we must deal with the costs of meeting particular sets of ambient quality standards.

Then the quesion is: How does the society ultimately choose among alternative possible ambient standards? Here is where we run into the full complexity of the political process. Total cost will be important, so our models can help at the simplest level by providing the so-called efficiency costs, to whomsoever they may accrue. But total cost, or the marginal cost of moving to higher quality, is not sufficient data for the collective (political) choice process. Distribution is also important - who pays, at least in the initial instance? Ultimately, then, if the management models are to be useful in informing political decisions, they must allow those responsible for the decisions to explore alternative environmental policies, especially as those are defined by alternative ambient quality constraints and alternative constraints on the distribution of the costs of meeting those constraints. It is important, we think, to understand that there is no biological (or economic) basis for choosing one level of ambient environmental quality over another.

DR. ORLOB: One of our problems, as professionals interested in some of the intricate details of ecologic modeling, is that we are not sure what, from the broader management context, might be the level of reliability required of the finished product. If someone can tell us what, in a regional economic framework, are the interesting levels of fluctuation in the shad migration, then we might be able to say that to get that resolution we shall have to model certain trophic levels in the ecosystem and to have certain environmental factors included. So, there is a problem of defining the range of interest at the outset. A lot of us are caught up in that.

DR. RUSSELL: Mr. Crook, would you address that? You are perhaps the only person here who is actually connected with a management agency.

MR. CROOK: I think the question itself is very pertinent. This is one of the things that we are trying to incorporate in our activities so we get a broad enough base to ask the proper questions in the proper terms before we get too far along in the modeling effort.

We don't collect data for scientific information, or for water quality purposes, before we understand what the public wants. That is, what they perceive are the things that they are going to get for the money that is going to be charged to them.

I want to make it brief at this time because it is part of my later discussion [of O'Neill's paper]. There has to be this communication backward and forward during the definition of the project and the setting up of the objectives.

DR. LACKEY: I would like to make an observation based on a previous comment. You were asking specifically about the kinds of problems we face in resource management. At least in our work at Virginia Tech-- primarily dealing with recreational fisheries--we have spent considerable time trying to extract information from the management agency about exactly what they want. We are not looking for a formal objective func- tion but just for some better way of defining the "output" from a fishery. Nearly everyone agrees that fish per se are only one small component of output, but it is difficult to identify the other factors management has in mind. This is, of course, a prior problem to that of measurement. Most agency people will say that they recognize this problem. But it is a long way from that realization to a set of uniform, accepted criteria which help to define whether a fishery is going to be better or worse after some specific management action is taken.

DR. ROBERTSON: I wonder where you are going to handle questions like the aesthetics. You mention the political arena, but I think of the way the Corps of Engineers builds a dam. The lobbyists for aesthetics have been very weak; they constitute a very diffuse public. Of course, the hope of some has been to give such factors some kind of economic value so they are included explicitly in the systems analysis. These seem preferable to leaving aesthetics to the political arena where their lobby is so weak.

I understand why you don't want to handle such problems. But, just throwing them out of modeling efforts bothers me. I wonder if you have some more explicit thoughts on how we are going to factor them in.

DR. RUSSELL: That is a good point. When we say the political arena, we are thinking of legislatures--either regional or national legislatures-- and are accepting that these bodies will accept responsibility for political decisions. This is simply not true now in this country. The Congress has instead created the EPA, and thrown political decisions to it, saying "You make the political decision, and if you do a bad job, we will call the Administrator on the carpet and chew him out." But EPA is not the appropriate place in which to make decisions about aesthetics and other intrinsically public goods (and bads). That, in our opinion is the reason aesthetics seem to be ignored. If we managed to dream up some way of including an aesthetic dimension for our models, we could perhaps obtain the satisfaction of having EPA develop pseudo-scientific standards for the dimension. But I personally am not convinced that would represent an improvement.

DR. DiTORO: Let us, for the sake of argument, suppose that the cost approach is the way to go. Do you really think that we have got the ecological models that make the kinds of optimizing calculations that you talk about worth doing? We have no inverse theory, so that even if you could say how accurately you needed to know the responses, we have no theory that tells us what kinds of precision and accuracies we would then need in our formulations. Without that the whole thing becomes judgmental. How do you know when something is good enough to set up large programming models and grind for hours on a big computer? We are an order of magnitude beyond our capabilities, in my estimation, in trying to make the kinds of calculations you seem to be talking about.

DR. RUSSELL: Your objection has the virtue of returning us to the subject of the symposium.

DR. ORLOB: I would like to go back to a technical point which, I think, may have some promise. It has to do with the penalty function and the marginal penalty term to which Walter (Spofford) referred.

In optimization problems, there exists a so-called "dual" of the problem which, in a sense, provides us the gradient on the response surface which we might examine to determine whether or not a marginal change will result in an improvement in the objective function value.

I think we need something like that even in ecologic models. We have never really come to grips with the complexity of the model in the sense of studying sensitivity and the interaction among variables. We really have no idea to what degree we might simplify the model by eliminating terms or modifying its structure in a direction which will keep us within the limits of reliability we have predetermined from the optimization problem. Perhaps there is an analogous kind of effort in the ecologic modeling area to that already formalized in linear programming and other optimization techniques.

Do you have any feelings for whether we should go in that direction? Will it help?

DR. SPOFFORD: I believe it will help. The marginal penalties, based on the penalty function and the environmental response matrix that I referred to in my presentation, are analogous to the "duals" of a constrained optimization problem. But rather than yielding the marginal increase in (real) costs of abatement for reducing residuals discharges, they provide the marginal increase in penalties associated with increasing discharges.

A relevant question, though, is, are we able to compute these ecosystem responses accurately, or are we one or two orders of magnitude off? In any event, the environmental response matrix, or the partial derivative matrix, is an approach to obtaining information on the environmental side that can be compared with information on the economic activity side.

DR. DiTORO: Suppose the model is plain wrong?

DR. SPOFFORD: If that is the case, and you realize it, you would certainly not rely very much on the model for management purposes.

DR. DiTORO: This is a likely situation, by the way. Many of these "researchy" sorts of models vary substantially. These models were just born three years ago.

DR. SPOFFORD: That kind of information would be very useful to us. If the ecosystem models of the type we have been discussing are, in fact, premature and only predict within an order, or two or three orders, of magnitude, this has implications for the amount of effort we should be putting into the development of management models that utilize ecosystem models as part of the broader decision framework. An assessment of the current state-of-the-art of ecosystem models is one of the objectives of this Symposium, and from our point of view would be one of the more useful outputs.

DR. ROBERTSON: I would like to carry on a little bit with what Dr. DiToro was saying. I basically agree with him; but I think he would also agree with me--since he is putting a great deal of effort into the development of models--that this approach shows great promise for the long term. I believe that we should at present develop and test models for real problem situations but not blindly accept what they tell us. We should take the output from these models to the people who have the best subjective judgement available--the environmental consultants--and ask them for an evaluation. Then we have, it seems to me, the best of both approaches to aid in solving immediate problems. For we have to make the management decisions now; we cannot say, "The models are not here yet; let us wait for 25 years." The decisions will not wait.

DR. SPOFFORD: There are other management decisions regarding
these models that we have to make today. Specifically, how much research
effort should we be putting into the development of aquatic ecosystem
models and how much, and what kinds of, data should we be collecting?
It would be extremely helpful to us if we knew just what it is that is
holding up the development of ecological models useful in making water
quality management decisions. Why aren't the existing ecosystem models
more useful for purposes of prediction than they appear to be? Is it
a lack of field data? Is it a lack of fundamental understanding of
the important interrelationships within the system? How much money
and effort should we be putting into model building and research on its
various components? What should we be doing now--in terms of ecological
research and data collection--so that in five or ten years we will be
in a better position to inform public policy on matters of water quality?

DR. ROBERTSON: I can give you an answer to that based on my
limited experience. What bothers me most at this point is the lack of
understanding of the functional relationships. I don't think we are
building the levels of sophistication into the models that are going
to take account of most of the things we should be worrying about. But
this is a tremendously expensive area. We should continue to support
it, but I don't think we should be too optimistic that we can quickly
develop accurate, predictive, environmental models.

DAY ONE - AFTERNOON

PARK, SCAVIA, CLESCERI PAPER: "The Lake George Model"
(Interruption related to a discussion of algae.)

DR. ORLOB: You have grams per square meter of water column. So, this is the total mass within the water column to a twenty meter depth, right? Have you looked at that in terms of the concentration of substances? How many milligrams per liter is that of blue-green algae? When do you see the collapse of the system by virtue of the production of blue-green algae?

MR. SCAVIA: We have not looked at it specifically in those terms. This is, however, one of the environmental perception parameters. We have made certain assumptions about where blue-greens really are and are not. The problem is that they are not distributed evenly.

DR. ORLOB: What happens to the oxygen distributed in that system when you get to that case? You are not predicting oxygen concentrations?

MR. SCAVIA: Right. We were told to discuss the shortcomings of the model, and one is that we do not model oxygen. This has not produced any problems with the Lake George simulations, because the Lake is always completely aerobic and never has any oxygen problems.

DR. ORLOB: But not with that kind of result?

MR. SCAVIA: Yes. That is one of the questions that we have.

* * *

DISCUSSANT: CHEN (see body of text)

* * *

(Interruption of Chen's discussion related to computation time.)

DR. THOMANN: What is the time integration step?

DR. CHEN: The integration step is one day.

DR. DiTORO: How many real variables?

DR. CHEN: Somewhere between 23 and 27.

DR. THOMANN: In 65 locations.

DR. DiTORO: Is there a differential equation for each one of those variables?

DR. CHEN: Yes. A one year simulation takes about 3.5 minutes of UNIVAC-1108 time, which according to the commercial rate, is about $35.

DR. ROBERTSON: I don't think you can get away with time steps like that [one day] if you are going to build in any kind of realistic hydrodynamics. You are going to put your price way up when you put in some reasonable hydrodynamics.

DR. CHEN: Yes. What I am saying is that it is going to cost more, but maybe that is the price we have to pay to analyze a complex ecosystem. I hope that people don't assume away the problem by saying we have to be cheap.

DR. ORLOB: One response to the question on hydrodynamics is that you don't have to have hydrodynamic simulation integrated within the quality program. It might be done separately and supplied as input to the quality model.

DR. ROBERTSON: We do the hydrodynamics in six minute steps and average over a twelve hour period. The averages are then added to the quality section. This saves us a lot of money.

(General discussion following Chen's remarks.)

DR. O'NEILL: CLEAN was designed primarily as a tool to guide and synthesize research. As such it should be considered supplementary to existing water resources models and not competitive. In my opinion, we should examine CLEAN for biological insights that can be extracted and incorporated into existing models designed for specific applications.

MR. SCAVIA: The direction that we are going with CLEAN--and I think it is probably a good direction to go if you are interested in biology--is to try to describe the system as complexly as we can. Then, through a kind of sensitivity analysis--where we play with the different terms in this very complex model--we find out just what kinds of things we can ignore. For example, we can find out whether excretion of phytoplankton is important. If it does not matter, take it out. We can explore the use of a very small time-step, and any number of other questions.

DR. ORLOB: I would like to ask a question that relates to the reasons for limiting CLEAN to just a biological system apparently without consideration of what I would think would be important environmental factors like dissolved oxygen and temperature constraints--both upper and lower bounds. Is that in fact planned for the future? Why not include oxygen within the model? I can hardly conceive of an aquatic ecosystem in which oxygen is not an important variable.

MR. SCAVIA: First of all, remember the context in which CLEAN was developed and remember it was implemented mainly for Lake George. That is why we didn't get involved with oxygen. Second, it is going to be included in the very near future because it is important, especially for trying to apply this model to other situations.

DR. CHEN: There is one more interesting twist to the dissolved oxygen problem. It appears when you apply the model to different systems. When we developed the Lake Washington model, dissolved oxygen did not go to zero at the bottom. When we applied the model to a Texas reservoir, we discovered that dissolved oxygen did go to zero at the bottom. As a result, when we went through the twenty years simulation,

nitrogen was building up. Because dissolved oxygen went to zero, the
model had to include denitrification in order to lose nitrogen.

Thus, if you don't include dissolved oxygen, you cannot include
bacteria because you do not know whether they are aerobic or anaerobic.

DR. CLESCERI: Gerry (Orlob), I think that Don's (Scavia) answer
was that the DO term is going to be included in the very near future
and is linked very closely to the decomposition sub-model where, in
fact, you have to switch from aerobic to anaerobic conditions and to
specify the products that are formed from decomposition. But, in
describing the DO regime per se we have not had it as a specific output.

MR. SCAVIA: The various terms which represent consumption and
production of oxygen are there; they have just never been collected.

DR. THOMANN: I think what I like about a session like this is that
you can really just lay it on the line. I will make a few introductory remarks, and then I will make a comment.

To me, a model is not a computer code. A model is not a set of
equations. We constantly talk about models in those terms. To me--
and, I think, to all of us in our shop--a model only becomes a useful
entity, for whatever purpose, whether biological insight, understanding,
or predictive management when the equations are shown in some way--and
this is a difficult task--to at least reasonably "represent" what you
are observing. Then the numbers that you put into those equations,
together with those equations and the entire structure, is, in our view,
a model.

That means that the burden is on the model builder and the analyst
to show that the equations he has written reasonably represent the
environment he is trying to simulate. I did not see that in the

Lake George presentation just now, and that militates almost immediately against transportability per se. You can transport code, but that is not transporting a model. A corollary is that the search for a general model is fruitless. Models will be site-specific. I don't think we have run a single model--even the simplest stream DO model--without making some changes.

So, how does one know he is improving his biological insight? It is not by generating output. It is not clear that because you change the secchi disc reading and something went up and something went down that that has improved your biological insight. Good biologists know that probably by intuition. You haven't told them anything new.

I really like the Lake George model because I think that, by and large, it has the right biological structure in it. It is a pretty classical structure, as Carl (Chen) points it out. It is true that I have not gotten any new insights into the behavior of Lake George from that structure. But what disturbs me most is the oversell of the model that I seem to get now.

The model has been put into a legislative framework and a public response framework. Even qualitatively how do you know that you are generating the "right" output? That seems to me to be pushing much too far with a model which has not yet in my opinion been shown to reproduce reasonably what you are observing.

MR. SCAVIA: You said a few sentences earlier that in our model, when we introduce a particular perturbation, the secchi disc reading decreases. You said any biologist knows that anyway, and that he is not being told anything new. I think that does show that the model is a model, since it is depicting what ought to happen, and that is one form of calibration.

DR. THOMANN: How do you know that? You haven't shown us that that is actually what happens in the lake.

DR. DiTORO: You have differential equations in there. How do you know they are right? I ask myself that all the time. How do I know, when I present a paper, that I am right?

MR. SCAVIA: I agree. You can get correct answers for the wrong reasons.

DR. DiTORO: But, how do you know?

DR. THOMANN: The only way you know is that what you are generating is observable in nature. If it is not, then you have not learned anything. You have only generated some output.

MR. SCAVIA: Why are you again saying that my simple examples are not observable?

DR. THOMANN: Because of where the data are. We have to see it. It is no trick nowadays to write down equations. The trick, in our opinion, is to write down the site-specific structure of the equations and to put the right numbers in them so that when you make predictions, they may come out.

MR. SCAVIA: Maybe I should have presented a little more in that area. As far as Lake George is concerned, our simulations are predicting what we are observing in Lake George--within an order of magnitude and with correct seasonal succession in the phytoplankton.

DR. THOMANN: I would like to see that.

MR. SCAVIA: There is a publication coming out very soon. (Simulation, August 1974)

DR. THOMANN: That is what the whole rest of the structure hinges on. How good a job have you done in representing Lake George's water quality?

MR. SCAVIA: I did not present verification results here because I thought the idea was to focus on the general formulation of the model and how it does or might fit into a management framework. I did not think I was here to show how the Lake George model works.[1/]

DR. THOMANN: Mine was a general comment. The Lake George model happens to be the culprit.

DR. CLESCERI: The other thing, too, is that linking it into management and the legislature is premature. The point is how the models can help--in the long run--in answering questions of management. I don't think the charge was necessarily to talk to ourselves again, but to go out on a limb and assume the model is going to get realistic output.

DR. THOMANN: That is a key assumption.

DR. CLESCERI: If you have a model that seems to work for a body of water, somehow you will go out and do some forecasting of how the body is going to look under certain assumptions about the future. Then some management scheme is inevitably going to be developed based on that. Similarly here, we have been asked by the Lake George people and other people in the Adirondacks to apply the model. We are not doing that yet--except at Lake George.

But we are going to interact with citizens groups, legislative bodies, various interest groups, and lobby groups which can be given greater understanding and insight by use of models. There is no question about that.

[1/] Editor's note. Illustrations showing the fit of calculated to observed data were added to the Park, et al. paper and are included in this volume.

DR. RUSSELL: I feel that I should take some responsibility for what will probably be a continuing source of comment and even tension. In talking to people who were asked to give papers about specific models, I told them to go easy on the mathematics, to go easy on the long tables and numbers, and to concentrate their 45 minute presentation and their paper on their problems. By implication I hoped that, if somebody had not been able to verify his model, he would tell us that that was one of his problems.

DR. NIHOUL: I agree with you that you must test the model against the data, but I would like to point out that the model may have another function. There is an enormous lack of data in most cases. A model may sometimes point out which data are most important--both in terms of what to measure and where to measure it.

For instance, there will be a joint data gathering exercise in the North Sea in 1976. All countries around the North Sea will take part.

I am the Chairman of a group called "Jonsmod" (Joint North Sea Modelling Group). We have been asked to provide advice and suggestions on the type and location of the measurements.

DR. DiTORO: I would like to know if you have dropped any equations yet? You said that one of the things that we were going to do with the thing was make it simpler. The last time I saw it, it was still very complicated.

MR. SCAVIA: It is a two-pronged attack. We are still making it more complex, but at the same time we have created a simpler lake eco-system model. The latter is a simplification of CLEAN based on what we have learned.

DR. DiTORO: Our experience has been that, even at the level at which Carl (Chen) is describing complications, they are just on the verge of not being understood at all.

I saw an application of CLEAN for Lake Wingra--obviously done by someone who did not know what was going on. I have never seen crazier looking calculated curves in all my life: limit cycles, oscillating to zero; all kinds of wild things, which you clearly expect from a set of thirty or forty ordinary differential equations such as you have. It is very difficult for me as an engineer to understand these things--or to even attempt detailed investigations of that sort of a model.

If it is a research model, then it is very important that you find out what does not matter and let us know quickly because we can go crazy otherwise.

DR. ROBERTSON: Our experience has been the same. You said that you wanted to continue to build in complexity until you got it as complex as you could and then simplify it.

I believe that we have to build relatively simple models, take a look at the results, and then try to evaluate what we can drop out and where we have to increase our complexity. Hopefully, such models will spotlight the important places in which increased complexity is needed. Of course, what part of the models should be improved will depend on the problem being attacked.

But this idea of just building in complexity and then simplifying it--I don't think you are going to be able to understand the results when you get the model "done." I think you have to weed out all along the way. And this is a challenge, I might say.

DR. KELLY: I think this comes down to a fundamental issue: That is, what questions are you asking of the model? If you want to evaluate the relationships between seventeen different species of fish, for instance, you cannot really build a model in which there is nothing except seventeen little boxes for fish. There have to be several other kinds of interactions.

DR. O'NEILL: Taking CLEAN as a totality and putting it in another model is like taking a carburetor out of a tank and trying to drive a sportscar with it, you don't want CLEAN: rather you want to extract the relevant ecological insights. Most of our present efforts involve analysis and validation of individual functions and interactions in the model.

DR. ROBERTSON: This is exactly the point I was trying to make. It is really going to be helpful if we can obtain analyses of separate components, for then we can fit in the parts we need for the specific problem we are trying to attack.

DR. O'NEILL: Our major interest is in designing and running rigorous experimental tests. If you compare model output to a large, expensive data set and it matches, you have learned very little, and you probably asked an uninteresting question. We will be seeking to invalidate the model, to discover where formulations do not correspond to reality so that improvements can be made. We are not seeking a data set to make the model "look good" but are striving to develop our understanding of ecosystems.

DR. CHEN: You learn more by making mistakes.

DR. DiTORO: How do you know when you make a mistake?

DR. CHEN: When it does not fit.

[Editor's note: At this point, Dr. Orlob discussed the stages of model development in a way closely paralleling that to be found in his paper. I have taken the liberty of leaving out the comments in the transcript. Only his final point is included.]

If validation fails, it can fail for one of three reasons. First, the data could be wrong. The second possibility is that the formulation might be wrong; and the third is that the parameterization might be wrong.

DR. LACKEY: This question is in relation with a previous comment. I am addressing it in a more general vein. This is an OMB kind of question--the kind a sponsoring agency might ask. I would like to address it to the discussant of the Lake George team or anybody else who is a potential sponsoring agency. What are the exact objectives in your type of work and, what performance criteria did you use to see if you were successful? The next question is: Have you been successful? How can you show the funder that he has gotten his money's worth?

DR. O'NEILL: The primary objective was to synthesize existing understanding and to use the models to draw implications based on the premise that the total system was too complex for simple intuition.

The test of the model is the search for counter-intuitive implications of initial hypotheses--and the test of those implications. This represents a rigorous test because the predictions you test are the ones you are unsure about. If the model predicts a lovely bloom of phytoplankton in the spring followed by a zooplankton peak, that is an uninteresting prediction to verify because the model system is doing what we know the real system does. What is intriguing is that under certain circumstances you get two phytoplankton peaks. Do we understand why? Is there a mechanism involved which we do not understand? That would be the implication that we would do research on.

That is an over-simplified answer because prediction is also an objective of the model--as it always is. But I think that our whole

scope of models is in very good shape, as far as empirical testing
of the individual pieces goes. As far as the total CLEAN model is
concerned, the next stage in the research is certainly intended to accomplish such testing, but we are not there yet.

I would be very surprised if it worked the first time we tried it.
I have no illusions about how much we understand about the system. To
say that I--or any team of us--can build a model of the biological components of an ecosystem to which you can give a rigorous test so that
you get the "right" answer straightaway, is to claim that we have a
wizard working for us with a magic wand.

DR. CHEN: In our case most of the people come to us because they
want to do something, and they need to have an environmental impact
analysis. If the project has not been built, we first have to show that
the model result is plausible. It has to be something with which you
can go to the public hearing. After we do that, we give them the program,
because the model is calibrated as well as we can for the specific
site. Finally, a model should be used on a continuous basis because
the users continue to get more data, and they continue to validate it.
They have the whole project to review. They have to have a post-project
monitoring program. So, they should continuously use that model for
management purposes.

MR. CROOK: I really think we are running past each other. Models
for research purposes are different from models for engineering applications and planning purposes. I think we are missing each other. I
think that is some of the problem in this discussion.

DAY ONE - AFTERNOON

KELLY PAPER: "The Delaware Estuary Model"

(Interruption concerning feeding function.)

DR. CHEN: I used the same function and found it to be wrong. I want to use a function which goes up and peaks at some optimum and then comes down. For example, for cold water fish, the feeding rate ought to peak at about 18°C. If the temperatures goes over 18°C the feeding drops down--in fact, it stops.

DR. KELLY: The behavior of all these equations coupled together is such that it gives a similar kind of behavior to that you describe. That is, there is a definite temperature optimum. In fact there is a different temperature optimum for different concentrations of oxygen. I am not sure whether that is correct, but at least, there is some indication that this can be so over limited ranges of temperature.

(Interruption concerning toxicity.)

DR. CHEN: The literature indicates that the influence of the toxicity level is coupled with the dissolved oxygen concentration measured in percent of saturation--not in absolute value. Do you amplify the toxicity effect when the percent of saturation of dissolved oxygen goes down?

DR. KELLY: No. With all the other errors I think this is a minor problem; but the point is well taken.

(Interruption concerning the second set of test data and the poor model fit to that data.)

DR. KELLY: What the model predicts and what is actually happening in the estuary are different enough so that I strongly suspect something is wrong.

DR. ORLOB: Is another hydrologic regime likely to be involved?

DR. KELLY: Yes. The second set of conditions involves about two and a half times the flow of the first set.

DR. ORLOB: If you compared the first set to another September with a comparable flow condition, would the model be hydrologically sensitive?

DR. KELLY: Yes, I think so, though that is a guess on my part. Of course, to get the output data and the input data for the same time in order to verify something is a real problem.

DR. ORLOB: Certainly.

DR. KELLY: To get the input data for this I almost felt like I was lying through my teeth. I took all the data together, and looked at the means. Then I compared it to what Dr. Thomann had in 1964 and I saw that the relationship between the two sets was very poor. In the center of the estuary there tended to be high discharges of BOD, and there tended to be smaller discharges on each end. That is where the similarity ended.

I did not spend a lot of time in trying to dig out hard data. I did not write to all the industries and ask them to tell me they were discharging. They did not tell the DRBC, so why should they tell me?

However, with all the problems in getting the values of the parameters within the model structure, the input data, and the quality of the flow data, there still appears to be something right. And there is something wrong as well.

Is the problem in the input data, in the structure of the model, or in the parameters used in the model? I have tried some sensitivity analysis, and found the results fairly sensitive to some of the parameters in the model. So, better estimates of those would at least give an indication whether there was something else wrong. But it would cost a tremendous amount to get that marginal increase in parameter accuracy.

Let us look at other aspects. This is essentially a pelagic model. There are marshes along the edge of the Estuary, especially down near the lower end. What role do these marshes play in determining the behavior of the plankton system? Then, too, let's say you have essentially one milligram per liter dissolved oxygen at the surface at a place where the water is about eight meters deep. The assumption is that it is vertically mixed. Down near the bottom the system has got to be anaerobic. How does that affect the way the rest of the system behaves? Is it important? What are the biota down there, and what are they doing?

So, here you have a model. It looks like it works. I ask, does it work? Do I believe it? And though I have many reservations, I say, Yes, in a general way. But I would certainly not want to stake my life on its accuracy.

DR. ORLOB: Would there be any value in comparing it to some of Bob's earlier runs with the model he has? There are some differences in the two. Certainly, your model represents a greater level of complexity but, after all, he did have a certain set of data and made certain runs that were credible. And then you could see where your model is particularly sensitive--if it is--to some of these parameters.

DR. KELLY: There is a problem with that. The inputs that I require are very different from Bob's (Thomann). If I have to try to verify my model on his data, that means that I have to make up the nitrogen and phosphorus inputs. Some of those data are available, and I can do it; but I cannot do the whole thing. In fact, I think the input specifications are different enough that you are really wasting your time trying to use one of these models to verify the other.

(Discussion of various technical points raised at the conclusion of the paper, prior to the formal discussion.)

DR. CHEN: You don't have a conservative substance to find out if the basic formulation in advection and diffusion is still valid, do you?

The reason I am asking is that we have a Mississippi river study that we approached very much as you have approached the Delaware. We calibrated for 1972--August, September, October, and so on. Then as we went along, we found that one of the months did not calibrate well. However, the conservative substance--we had a chloride concentration-- showed the physical processes to be correct. Then, we found out that there was a temperature effect that was out of whack. Within a certain range our assumptions worked, but when we had a drastic change in temperature the assumptions broke down.

DR. KELLY: For the May 1970 data the temperature is ten degrees lower.

DR. CHEN: It is ten degrees lower. That raises again an earlier question on the temperature response functions.

DR. KELLY: But, again, you were talking about an individual species. When you are aggregating your data, what difference does it make? Does that response change? That is a question for ecological research that I would like to see investigated.

DR. ORLOB: What is your particular rationale for using any particular set of dispersion coefficients, other than the fact that Dr. Thomann provided them? Are they universally applicable? Obviously, it looks like you could make that curve go either way by tinkering with these coefficients.

DR. THOMANN: The original coefficient was derived from the verification with the time variable on salinity.

DR. ORLOB: Inasmuch as there is no conservative constituent in the Kelly model, there is no way of checking whether it is, in fact, consistent.

DR. KELLY: I did run the model with sea water boundary conditions at the end to see what would happen. The profile was approximately correct. Let us look at this, too. Is it reasonable to assume that the fish concentration approaches one milligram per liter or more in the Delaware? I don't think it is. The same thing goes for the bacteria and zooplankton. The concentration of algae does not look bad.

DR. ORLOB: It is possible that algal blooms are visible at those concentrations? Have you ever seen algal blooms in the lower end of the estuary? You should be able to see blooms at biomass levels above about 1.5 mg/ℓ. This has been our experience with Lake Washington.

DR. RUSSELL: What about the fish? Can anybody translate milligrams per liter of fish into more meaningful terms?

DR. CHEN: Fish biomass ought to be expressed through kilograms per hectare (or pounds per acre). Most empirical data are expressed in those units. If your prediction is several hundred or a thousand, you know you are out of whack.

DR. RUSSELL: Then there are data once you make that translation? [from mg/ℓ to kg/ha]?

DR. CHEN: That is correct.

DR. ORLOB: Do you allow your fish to migrate to where the food is?

DR. KELLY: Yes. They are not controlled by advective terms.

DR. ORLOB: So, they may either vacate an area or move into an area where there is something that attracts them?

DR. SCHAAF: But you are assuming that the food limit is upstream and downstream, not in the middle?

DR. KELLY: I am not assuming that; that is what the model says. If the model's structure and the coefficients and the data input are to be believed, then the fish apparently are food limited at the top, food limited at the bottom, and oxygen limited in the middle.

DR. NIHOUL: On what basis did you choose your state variables? Was it an educated guess? Was it because most models do it that way? Or did you have a series of data and find evidence for the choice in these data?

DR. KELLY: I knew what the inputs to be managed were; that is, I knew they had to be consistent with the industrial and municipal treatment plant models. I also knew we wanted outputs meaningful to people and to policy making. The rest of the model comes from my intuition about how the system might behave.

DR. DiTORO: How did you compute BOD for the BOD verification? That is not a state variable. Is it just bacterial respiration?

DR. KELLY: No. BOD is a state variable.

DR. DiTORO: Oh, it is? So, you are grinding it out. When you computed BOD verification, did the bacteria play a role in that calculation at all?

DR. KELLY: You raise a good point. BOD is really the sum of all the organic materials in the Estuary. So, actual BOD measurements reflect zooplankton, bacteria, and algae, too. Initially, I ignored that. Then I saw that the output did not look very promising, so I asked myself what would happen if I changed the model to compute BOD as the sum of all the biotic components that we can measure? I worked on that for about two months, but it never came through.

DR. CHEN: I have a couple of comments. In the presentation to a public hearing on a fish model, biologists are sitting there ready to clobber you. They will point out that your fish production computation is not meaningful with respect to other environmental factors such as the suitability of the spawning ground and its availability. The model cannot possibly address itself to that kind of question. So, we limit ourselves to saying that BOD is bad in that low DO decreases the growth or increases the mortality. This does not mean that if you clean up the BOD you will have more fish production. That is one point.

The second point is that in an estuary such as the one we are looking at, the presence of migratory fish is going to be correlated to the salinity distribution, the temperature distribution, and flow - rate. To come back to my own experience, we did a study to find out the determinants of migratory fish survival and it turned out to be independent of dissolved oxygen or whatever was in the system. (I can speak only of one specific system in which the dissolved oxygen never quite went to zero.)

DR. KELLY: In partial answer to this, there used to be a very large shad migration in the Delaware. When hearings on Estuary water quality standards were held, it was maintained that the major obstacle to the migration of shad was the oxygen block in the center of the Estuary. Part of the reason for choosing the water quality standards that were adopted was that at certain times of the year higher levels of oxygen had to be maintained throughout the estuary than at other times to remove this obstacle. This model does not address that time-varying aspect of the management problem.

Including migratory fish explicitly would mean expanding the model greatly. You could say I expanded this model much past the data base already. So I don't know how much I would gain by going on to include the migration of fish in the system.

DR. CHEN: We did it with an entirely different approach. We just addressed ourselves directly to the migratory fish.

DR. WOLFE: I would like to make one comment in relation to this. The comments on migratory behavior so far have been oriented towards adult fish. It may be that the larval fish, which you have not considered as a state variable, do play a very important role in influencing the sizes of some of the compartments that you have looked at. Gordon Thayer at our laboratory has data that suggest that consumption of zooplankton by the larval fish in the springtime may, in fact, be responsible for diminishing the zooplankton peak populations which follow the well-documented phytoplankton peak.

Now, larval fish probably enter the Delaware Estuary in late winter or early spring and will orient very specifically towards other environmental parameters like salinity. So, within the reaches of your system, larval fish could be accounting for some of the discrepancies between predicted and observed values during May and September, particularly in the low salinity portions of your estuary.

* * *

DISCUSSANT: DiTORO

* * *

DR. DiTORO: I would like to talk about what I went through when I saw Dr. Kelly's paper. I tried to decide whether or not I believed the calculations; whether they are reasonably plausible; and what stance I would take relative to those calculations if I had to comment on them in a professional way for use in water quality management. In order to tell you how I have done that, I want to spend a couple of minutes giving you the point of view that, in fact, our group adopts relative to these kinds of calculations. It is as follows.

There are two major effects that are critical to modeling micro-organisms up through the zooplankton. They are both equally important. They are: one, the transport structure of the body of water and, two, the kinetics of the microorganisms involved. The transport, once set by observed data or hydrodynamical calculations, is sacrosanct. You cannot fool around with it. Other people have done transport calculations on the same body of water. Yours had better check theirs, or find out why they don't.

Secondly, micro-organism kinetics are the same the world over. Algae in Lake Washington behave more or less like algae in Lake Erie--with a certain range. That range is beginning to be known. Enough phytoplankton models have been built so that the kinds of coefficients that go into the formulations fall within certain ranges. Those ranges check kinetic experiments as well as prototype data. We have a lot of these experiences for BOD-DO models. When one checks a BOD-DO model, one knows what values the kinetic coefficients ought to have. If they are different, one has to see an explanation. If no explanation is provided, one pays no attention to the model thereafter. The calculations are just wrong.

I cannot stress this point of view enough. Algae behave the same the world over. The kinetic formulations are the same. What makes each lake different is its physical mechanisms--the mixing, and so on-- its nutrient levels. The way the micro-organisms behave is invariant from lake to lake. That is a somewhat controversial point among biologists. They have been taught--and continue to believe--that every lake is peculiar unto itself. But if it is true that the fundamental structure of micro-organisms are different from lake to lake, then any generalizable theory is forever doomed.

Now, that does not contradict Dr. Thomann's comment about the fact that a model must be a lake-specific. What he is talking about is that the parameters which are lake-specific--extinction coefficients, dispersions, loadings, and so on--must be done for that lake. But if the fundamental structures have to change lake to lake, you are in big trouble. It suggests you don't know what is going on.

With that prelude, how does one judge if a particular model is any good or not? Well, the first thing that I looked at was the figures for which Bob (Kelly) had some data. Does he at least duplicate in his validation set what was observed under the same conditions? He does reasonably well as he showed you on that one figure.

In any case, if one has no data, what does one say? My view is that, if you don't have any data, you don't know what is going on. Calculating numbers, which you have not verified in some way, is extremely dangerous. Therefore, I immediately wrote off the toxicity and the fish calculations. I am a fan of including bacteria in water quality models. I do that without telling anybody. But they looked very high--two milligrams per liter of dry weight is a lot of bacteria. The zooplankton was also very, very high--almost as much zooplankton biomass as phytoplankton biomass. Those observations made me rather skeptical and I would have rejected the output not validated against data. I would have done this primarily because the order of magnitude wasn't right.

With respect to the variables to which you are paying some attention, is the structure reasonable? First, you are using conservation of mass equations. This is absolutely essential in my view. Anyone who is going to build a model must at least start with something that is certain. Namely, all the mass has to be accounted for in one way or another. Carl (Chen) would have never found he needed denitrification if his equations didn't conserve mass. I can cite numerous examples where that kind of effect shows up because you are sure of something. If you do a model in which everything is an hypothesis and in which all reaction coefficients are hypothetical or picked-to-fit, then you are never surprised because you don't know anything.

Do you do the transport reasonably? There was not any dispersion in the early versions of the model. Anyone who has been brought up with BOD-DO as a background is immediately suspicious. But, it is possible to calculate the so-called estuary number, or transport reaction kinetic number, to give you an indication as to whether dispersion should be important. I did so for your situation and it said that for those flow regimes in the upstream region you probably were marginally correct but you were incorrect in the downstream regime. And, of course, the salt concentration tells you that, too. So, there was

a problem. But you could make some guesses about what might be the
effect.

I move now to kinetics. Here is where it gets sticky. The reaction kinetics that I am used to are based solidly in laboratory numbers.
There are ranges, but everybody knows that saturated growth rate of
phytoplankton, for example, at 20°C, has to be between 1.5 and 2.5 per
day--at optimum nutrient and optimum light conditions. There is just
no getting around it. You look at the numbers, and if that reaction
coefficient isn't between 1.5 and 2.5 per day at 20°C, that is the end
of the game. You ask yourself, Why isn't that right? Never mind the
rest of the model. And so it goes, constant by constant, wherever
these constants are available.

Now, if you have a formulation where the constants have no experimentally verifiable counterpart, you have conjectural parameters. If
you have thirty conjectural parameters in your formulation, there is
no way of checking it. Then, it becomes an elaborate curve-fit, in
my opinion.

Are the light and temperature effects reasonably known to an
order of magnitude? Some are and some are not, as Bob (Kelly) well
knows.

What do the kinetics do without transport? Do they reproduce observed laboratory behavior? With a very complicated model that is very
hard to say. At least, for phytoplankton you can calculate nutrient
kinetics and convince yourself that, in fact, they work very well.
We have done this. The more interesting question is linear versus nonlinear kinetics. As some of you may know, the classical theory of BOD

and DO is based solidly on linear kinetics. The most surprising thing
in all of my experience has been that it works as well as it does. It
is one of the mysteries of our business, I would say; and a paramount
question which has to be answered in order that we get to a more
fundamental   understanding, is why it works. Bob (Kelly) did not
use linear kinetics for BOD-DO. He used bacteria and actually built in
the bacterial phenomenon. I think that is praiseworthy, but it has
got to work. This is why I asked that question about how he compared
the BOD to the measurements. I think his comparisons are done
incorrectly.

Looking further, you come to the projections. Do the projections
look reasonable? If I see things that look screwy--"screwy" is a
loaded word, and I will try to tell you what I mean by "screwy" in a
minute--then I start having to find out why they did what they did.
Until you can convince yourself in some plausible and rigorous way that
you understand the results, you have a model which is as mysterious to you
as the set of data is. If you don't understand the intricacies of
your own calculations to the point where you understand every wiggle,
what has it gained you to calculate? You are reduced to saying, "First
the model did this; and then I changed this; and then the model did
that." So, the projections have to look reasonable. They have to be
explainable in terms of how one thinks these equations behave.

A technical point: are the computational algorithms reasonable?
Is the numerical analysis done correctly? Anybody who has tried to
integrate large sets of non-linear, partial, differential equations
knows it is asking for trouble. The hydrodynamicists have three, four,

or five equations, which are very fancy but known equations. There
is no question that those hydrodynamical equations describe all the
fluid flow. And yet, they cannot integrate those things in any
reasonable way. All of hydrodynamics for the last 100 years has been
trying to solve the differential equations that have been known since
Stokes.

You have to have at least some faith that your numerical analysis
is reasonable. There is a whole history of literature over the past
couple of years in which people have been arguing about the size
of the dispersion coefficient in tidal estuaries. It turns out that
the whole argument was really about what kinds of numerical errors
were being made by the various integration schemes that were used to
calculate these solutions with the tidal velocities included. This is
a most embarrassing episode for all of us.

So, you have a series of non-linear equations whose properties are
poorly understood, and you are trying to integrate them. How do you
check numerical algorithms? You would like to check them against
analytical theory. But there is no analytical theory in this business.
(I will get back to that.) In the case of the simple BOD-DO equations,
on the other hand, there is analytical theory available. The system
is described by two linear, partial differential equations.

The analytical theory of those equations is understood reasonably
well by certain segments of the profession. It provides you with a
framework, by telling you how those differential equations behave. All
kinds of peculiar things come out of that theory--effects you would
never know about had you not seen the analytical theory. These include:

boundary condition effects, flow versus reaction, how loads affect
calculation, how reaeration works, and so on.

There have also been many applications of the numerical models--
the prototypes--all of which look reasonably good. Once you have done
ten of these, you begin to believe in them. People have predicted what
was _going_ to happen. And they didn't do badly. The Thames Estuary
is the best one; the Mohawk another. On the latter, quality conditions
were predicted under the assumption that 85% removal of carbonaceous BOD
was required. Then they went out after five years, measured what had
happened, and found that the predictions were fulfilled. If you get
a few of those under your belt, you become a real believer.

What is the situation with the phytoplankton biomass models? There
is almost no analytical theory available. Even for very simple sets of
equations, nobody really knows what the solutions are. One of the things
I personally am doing is trying to get some analytical solutions to
see how these models ought to behave, at least, in portions of the
response space.

What kind of backup do we have on kinetics? At last, the kinetics
are being done systematically. Phytoplankton researchers have seen the
light, and so Michaelis constants assimilation rates, and stoichiometric
conversions are being measured on a lot of different species.

One big lack **is** respiration. Nobody knows what phytoplankton do
in the dark. Yet, about half the biomass is in the dark, and the
kinetics are implying for the computation what the phytoplankton are
in fact doing in the dark. But there is very little hard evidence on
which to check these implications. Nonetheless, the verifications on

phytoplankton models are beginning to be convincing. Dr. O'Connor
will show you some things we have done which will give you an idea of
where we are relative to phytoplankton model verifications. These are
simple phytoplankton models. What we play with are phytoplankton,
nutrients, and zooplankton. The zooplankton is in there, not because
we like it but because we had to put them in. We refused to put
anything else in unless someone could come up with a really good reason.

There are as yet no cases of successful prediction. For example,
cases in which someone has said, "You take out 85% of phosphorus at the
Blue Plains Treatment Plant and the Potomac will clear up." And then
been able to verify the prediction by measurement.

The models for fish and benthic interactions are another order of
magnitude more complex. If we apply the criteria I suggested at the
beginning, they too have to be independently checked. In that connection
I would be interested in how seriously Carl (Chen) takes his fish biomass
calculations other than to check kilograms per hectare and say that at
least it is on the same piece of paper as some measurement from a reservoir
two hundred miles down the road. I suspect that is about all the fish
models are worth.

DR. CHEN: That is, in fact, the case.

DR. DiTORO: For anyone to say that you can tell what species is
going to come in or go out if I change the phenol concentration in a
plant effluent on the Delaware is just so far beyond existing capability
that it is laughable.

What have we really got, then, with the Delaware model? I submit
to you that what we have is a BOD-DO model with a little bit of nitrogen

done with bacteria--which is not to be sneezed at since I think that
is the way modeling ought to go. But that is all.

DR. ROBERTSON: Regarding the problem of the bacteria and zooplankton looking very high; first of all, I have seen a number of
studies in which the zooplankton biomass is about equal to the phytoplankton biomass. A number of Polish lakes have shown this result,
so it is not completely out of the question, although it is somehwat
surprising.

Secondly, I wonder if we are fooling ourselves
As I remember the equations we were really looking at heterotrophs. A
number of the smaller nannoplankton might be functioning as heterotrophs.
If we added this fraction of the nannoplankton to the bacteria in that
estuary, you might get a biomass at the level seen in the model.

DR. DiTORO: It is conceivable. Heterotrophic nannoplankton
simulation is an uncertain business. With light available, they behave
like phytoplankton.

DR. ROBERTSON: But what is your turbidity there?

DR. DiTORO: We don't have the measurements.

DR. ROBERTSON: What I am saying is that the bacteria result should
not be thrown out immediately as wrong, though I am certainly not
willing to claim that such high values are actually valid.

Finally, I would like to ask you for verification of your statement
that algae in all lakes function about the same way.

DR. DiTORO: Our experience has been that the kinetic constants
tend to fall in a relatively narrow band, except in one interesting
example where we knew that we did not have a very good model. For
example, the rate at which BOD goes away is somewhere between one tenth

and two tenths or 0.5. But, you must have the numbers.

DR. ROBERTSON: I would like to see more proof than I have seen that blue-greens in a farm pond are going to have the same general kinetics as diatoms in Lake Superior.

DR. DiTORO: I agree, but the fundamental hypothesis that under-lies all of this model building--at least in the way we practice it--is that in biomass terms and within the limits of precision you can get, such universality holds.

DR. CHEN: I have a pet theory about why BOD works the way it does. My theory is that the key bacteria is of very small size and can saturate its nutrient need for growth very fast compared to bigger ones.

DR. DiTORO: Yes. But what is wrong and what is missing is that the growth rate of bacteria is known to be faster than algae by almost an order of magnitude. So, why do linear kinetics work? I have never been able to come up with a convincing theoretical solution. I leave that to the students.

DR. CLESCERI: You mentioned bacteria in the lab. They are not necessarily the same in the water.

DR. DiTORO: Well, the argument is that the bacteria are the same the world over.

DR. CLESCERI: That is not what I mean. My point is that you made a comment that bacteria grow faster than algae.

DR. DiTORO: Yes. Saturated growth rate.

DR. CLESCERI: Yes, but that is a laboratory observation.

DR. DiTORO: Yes, certainly.

DR. CLESCERI: My comment is that it need not obtain in a lake or a stream.

DR. DiTORO: But I am saying it does. I refuse that as an explanation. Bacteria are the same the world over--just like algae.

DR. CLESCERI: I think that is the basic error here: trying to extrapolate from a lab experiment to the field. That is one of the reasons we are where we are. For too many years, we have tried to devise experiments in the lab that mimic the field. It is very diffi-cult to do that.

DR. THOMANN: What we are really trying to aim for is to put what we do on as firm a scientific foundation as possible. This means that the people in this room will ultimately have to give our work a peer review. Then they will be asked to agree that our hypotheses are reasonable because they are based on such and such data in the labora-tory or field. There are a whole series of verifications, but if we don't do this, we are all going off in our own direction.

DR. DiTORO: All a bacterium knows is what is outside its cell wall. It does not know whether it is in a chemostat or in a lake. Is your fundamental argument that the bacteria somehow knows that, when it is in a laboratory, it should behave differently than when it is in the field?

DR. CLESCERI: Because observed growth rates have been different in the field than in the laboratory.

DR. DiTORO: Yes, because there are other variables.

DR. CLESCERI: Competition, for one.

DR. DiTORO: Exactly. But if you structure a model without the competition effects and other similar complications the bacteria had better grow according to the laboratory growth rates; otherwise the whole thing is ridiculous. Isn't that reasonable?

DR. ORLOB: I think you are going to find that the experience which has been confirmed in BOD-DO work is going to be confirmed in algal growth in the field--ultimately. We are close to that now.

DR. DiTORO: It is very close.

DR. ORLOB: As we go along a little farther, it will be confirmed with zooplankton characteristics in the field--if we can get the data for that. I think we are still a little way from that now.

DR. RUSSELL: So, far, all the talk has been about parameters. You measure them in the laboratory, and then you stick them into certain functional forms. But I don't see how you can possibly know that those are the appropriate forms. I am concerned that a lot of effort is going into argument about and measurement of parameters when you may not even be using the right functions. How do you know one way or the other?

DR. DiTORO: We have faced exactly that problem, and so has Carl (Chen). Our solution was the following: we went into the literature, found a set of beautiful experiments by Ketchum in the 1930's, and checked whether or not a product of an inorganic nitrogen Michaelis function and an inorganic phosphorus Michaelis function fit that data. And it did. That, at least, gives you some faith that you are not kidding yourself.

DAY TWO - MORNING

NIHOUL PAPER: "The North Sea Model"

(Interruption related to bacteria and other data)

MR. SCAVIA: In reference to the slides of the bacteria and the different parts of the food chain--were those observed or predicted?

DR. NIHOUL: Those are the observed values.

(Interruption related to management use of the North Sea Model.)

DR. HARRISS: I am concerned about the use of this type of model for setting criteria for making a decision on ocean dumping. Taking the case of mercury, as an example, it seems to me that the real problem is that we still do not understand the mechanisms involved in mercury toxicity--particularly the sub-lethal effects which, I think, are the real problems, not the acute effects where you go out and dump tons of mercury.

For example, look at Mobile (Alabama) Bay, which has been highly contaminated but to which the major mercury inputs have been cut off. It turns out that there are some very subtle problems related to the circulation of mercury in the Bay. These are related to short-term phenomena. For example, the mercury that is now in the sediment, which is an anaerobic environment, is released during periods of high wind stress producing pulses of high mercury contamination. It turns out that most of this mercury is not methelated but is associated with a natural humic acid compound. It is only toxic--as far as we can tell--to about ten percent of the phytoplankton species in the bay. Thus, we cannot lump all the plankton together and look at the toxicity of mercury to "plankton" because we face selective hitting of certain species with certain mercury compounds in question.

DR. NIHOUL: As we said yesterday, the determinant is the question you are asked to answer.

If you are concerned about the effect of a particular form of some heavy metal on a particular species of plankton, you need a model which can describe that very specific interaction. If you are asked to evaluate the transfer of heavy metals through the food chain and relate inputs into a given marine region and ultimate concentrations in fish catch, then you cannot go into the details of the innumerable interactions. What you need is a compartment model predicting the translocation of the pollutant from water and sediments to phytoplankton, to zooplankton and finally to fish. The objective is to provide the economists with a transfer function relating the inputs--the amounts and locations of which can be controlled in the economic context-- to the concentrations (in fish for instance) which are controlled by national or international regulations.

DR. KNEESE: We didn't get a very clear impression of just where the model stands. Are you entirely finished? Or, are you still elaborating in the model? What kind of schedule are you on?

DR. NIHOUL: In our latest report to I.C.E.S., you will find the result of what we had found up to December 1973.[1] At that time, we had detailed hydrodynamic models but we had only budget models, - using yearly averages - for the three regions which the residual circulation pattern had revealed.

---

[1] Math Modelsea (1974), Mathematical Models of Continental Seas: Dynamic Processes in the Southern Bight, I.C.E.S. Hydrography Committee, C.M. 1974-C:1

We have developed, since then, time dependent box models, i.e., differential models giving the yearly variations of the concentrations in the three zones.

These models have already been validated and tested in the inner lagoon region - the sea pond near Ostend, Belgium. They fit quite well there, but they have to be validated for the three regions.

DR. CHEN: Are your bacteria measured in water column or in the sediment?

DR. NIHOUL: Both. These pictures show the bacteria in the water column.

* * *

DISCUSSANT: QUINLAN

* * *

MS. QUINLAN: I could bring my technical background into the discussion in great detail and look at the ecosystem processes and at the dynamic processes. But I think we have heard enough of that. I believe we should re-focus on the purpose of the conference, which is ecological models in a resource management framework.

There are a lot of people here, whom I don't know, who are probably more in the management domain. I would like to know what their attitude is toward the information that comes out of these physical or engineering models or the more functional biological models. Is it in a useful format? There are decisions that have to be made. There are questions that one must ask in order to make decisions. We, in the engineering community, try to structure our models according to certain physical laws or principles to answer these questions. What about

managers? Has this information--or can this information--be transferred into the management domain? I would like someone to address
himself to those questions.

DR. CHEN: I think that the models have been used only on a
limited scale because they have been opposed by the managers. The applications which have occurred probably did not result in answers to all
the pertinent questions to everybody's satisfaction, but, if somebody
were to ask, for example, What will happen if we do this? Would the
DO go bad? Would the toxicity increase? On this limited scale, they
have applied and, in my judgement applied successfully.

MS. QUINLAN: But you are a person who generates this information.
Can someone from the other side say how useful this sort of information
has been?

MR. CROOK: I think we have to be more specific on this. What
type of information are you talking about? Is it general ecological
modeling, or do you want us to answer something specific? I think it
makes a lot of difference.

DR. ROBERTSON: One of the problems here is that resource managers
are not represented here. I think Leonard (Crook) is halfway there,
but I don't think there is a decision-maker here. At least, the people
I know are not.

DR. ORLOB: That raises a very interesting question. Who are
these "managers"? Where are they?

DR. ROBERTSON: Sometimes they are very difficult to identify.

DR. CHEN: I think I answered the question on the wrong wave length.
Let me try again. In a pumped storage study our model study showed that

they ought to deepen their tunnel. It cost them about $300,000 to re-design the whole thing, and the construction costs were very high--the estimates are not out yet. But they did accept our recommendation and went ahead to spend another $300,000.

MR. CROOK: And your model was based on ecological considerations?

DR. CHEN: Yes.

MR. CROOK: That is a partial answer for the design context. In a planning context we are trying to move--and very cautiously I might add. We have been at this for about five years, as some of you know. I think Andy (Robertson) started with it in 1968. We have been trying to test the water so to speak; trying to find out just what you modelers know--and what you can give us. We contracted with Hydroscience for a look at the state of the art, to tell us what we could rely on, and how feasible it would be for us to do a specific operation, such as regional planning, for the great lakes, in certain areas. They told us that we could do it in certain areas with the data we had and with the state of the modeling art. We are now proceeding to get funding to do that.

My answer then has to be a very strong yes. We are looking for results from ecological modeling. We want--and this is part of my dis-cussion, but I don't mind exposing it because of the question--to work very closely with the modelers, to specify the objectives of the study very precisely; and keep in constant touch with the modelers and the scientists in order to get the answers we want at the minimum cost in a minimum time. We see these models as the only source of the answers we need to allocate resources in the Great Lakes.

DR. DONALD O'CONNOR: There has been a history, at least over the last twenty or thirty years, of the first steps in ecological modeling. These early models have been used and are currently being used by the EPA in the assignment of their limited funds to the very significant problems they have. Our classic DO models for streams and estuaries are now being very, very heavily used and--I am pleased to report-- even accepted by policy makers. They are beginning to believe in models. In that sense, although it is the first step in the ecological chain, those models have had a long history of acceptance. I want to reserve my future remarks for my own talk.

The next level of modeling is now being accepted and, I think, with people like ourselves accepting the responsibility for communicating with the administrator, I am rather optimistic about their use.

DR. RUSSELL: I suspect the answer to the question of utility and acceptance is yes and no. The technically trained man likely to be found at the head of an executive agency--or perhaps high on the staff of a commission, may find the models useful while understanding their limitations. The lay person, for example a legislator, is likely to find them mystifying and to be unable or unwilling to make sense of the output (fish in mg/l, for example).

MR. CROOK: I think that is perfectly valid, but we don't want to let Alician (Quinlan) off the hook here. We want to hear her discussion of this. I think she has tried to spark an interest and wake us up. Andy (Robertson), would you like to add to that a little bit?

DR. ROBERTSON. I have a comment on the suggestion that the models don't affect legislators enough. My experience has been that legislators usually have hired experts. They have a group of in-house experts. If you can convince these people with your model, then they can often convince the legislators.

DR. ORLOB: One of the things that has changed in planning and engineering in the last ten years--perhaps largely as a result of the capability represented by the computer and the whole area of mathematical modeling, including ecological modeling--is the attitude of engineers. In the past engineers made all prior decisions and judgments on what might be done and thus (allegedly) imposed their value judgments on society. That practice has changed toward introducing more alternatives for consideration. Therein, I think, lies the value of the models. The alternatives considered today are, in general, very complex. Most often, they involve major alterations of the environment for which there is no precedent.

One of the things that we need to be able to do then, is to evaluate the differences between alternative strategies that we may impose on the environment. In this regard, I think a model, even though it may have some limitations, may provide us with a useful tool for assessment of the relative consequences of alternative courses of action. Perhaps we should be looking at the differences that are manifest in these alternatives with more attention than the absolute values predicted for any one particular course of action.

The communication we have to effect, I think, is between the model technologist and the next level of management. That is, the people who are also informed technically and who carry that message forward in the form of a wider choice of alternatives to the decision-makers.

I do not believe we should delude ourselves into thinking that we, as modelers, are going to communicate directly with the decision-makers and convince them that our product is something worthwhile.

DR. NIHOUL: I agree that there is a problem in evaluating the social benefit. However, in most cases we don't need to. For instance, there is an international regulation on the amount of mercury which can be consumed. I don't know whether this value is correct,nor if there is any social benefit to it, but there is a regulation and the managers cannot allow more than a certain concentration of mercury in the fish. There is nothing between the existing political decision and the model.

DR. THOMANN: I would like to dispute the claim that we are not generating output that is relevant to decisions. Our experience over the last twenty-five years indicates otherwise. Over that twenty-five year period there have been numerous examples of the application of the very simple BOD-DO equations. The output of those applications has been dissolved oxygen. Nobody in any kind of decision-making position--let alone the man in the street--really "understands" this output. And yet, those models have been effective over the twenty-five year period and have influenced decision-making successfully, as we tried to point out yesterday.

Why is that? Dissolved oxygen is a chemical output that we really don't understand. I believe the reason these models were taken seriously is that the scientific community as a whole agreed that the models, as they were being used, represented at least a valid starting point. Then, there was a translation of the chemical quality to a relevant variable; in most cases, by inference. If the DO drops too low, fish are going to be harmed.

How does that apply to the phytoplankton model today? The phyto-plankton models are only three or four years old. They just have not been around long enough for a scientific consensus to develop. But our task is similar. The models have phytoplankton biomass as principal output. Somehow we have got to translate that into a meaningful variable for a decision-maker. For example: "In the backwater you are going to get unsightly conditions" or "concentrations above 100 micrograms per liter will produce an undesirable condition." We are still, I think, searching for that level.

DR. KNEESE: The usefulness of models of natural systems--wherever they are-- is very much dependent on the policy context which exists at a given time. When Bob (Thomann) started his work on the Delaware Estuary, we had laws at the national level which encouraged work of that kind. It went off to a flying start.

It was a classic study which used such models and tried--I think very successfully--to integrate them into an ongoing decision-making process. But, now we have had a series of laws, at the national level at least, which are kinds of know-nothing legislation. They say that we want to make decisions without knowing anything about the environment.

DR. RUSSELL: I am playing the role of the devil's advocate in this discussion because I thought I heard too much self-congratulation in the initial response to Ms. Quinlan's question. I am certainly sympathetic to this kind of enterprise.

I think the example that Allen (Kneese) tacked onto the end of Bob's (Thomann) comment is perhaps a good one to probe a little more deeply.

In reading what I have of the Delaware hearings, it has struck me that the weight of the water quality decision for the Estuary rests on the shad calculation, which in turn struck me as made up out of whole cloth. Very little of the weight of the decision appeared to rest on the DO model--however good it was.

I would suggest that, though in the past 25 years DO models have been built and run and decisions have been made, it is far from clear that the link between the two sets of events is as strong as you think.

DR. THOMANN: I don't believe that. I could provide examples from all around the country. There were the classic studies on the Mohawk, on the Ohio--even before the original Delaware study. There was also a study on the Hudson. The degree to which the decisions were actually influenced by the models may be debatable. But this recalls Gerry's (Orlob) point: in the final analysis anything we produce is only one input to the decision. We cannot hope for more than that.

DR. KNEESE: I think there is even another dimension to the whole thing that we haven't talked about much. If I understand correctly, most of these are steady-state models.

MS. QUINLAN: No, they are not.

DR. O'CONNOR: A lot of the DO models we are talking about are.

DR. KNEESE: Would someone like to indicate how probability enters into this? Can the models be used to predict not only what the population of fish would be, but what the probability distribution around that point prediction looks like?

DR. THOMANN: Let me point out that now you cannot build a model which predicts, even deterministically, the fish biomass with any degree of accuracy.

DR. KNEESE: Even if you could predict deterministically, you would still have the other aspect of the problem.

DR. THOMANN: By and large, stochastic elements have not been dealt with.

DR. KNEESE: They may, however, represent an equally important part of the decision problem.

MR. SCAVIA: What you are talking about is the probability of your decision being correct. Before you estimate that, you have to do what the DO modelers have done; that is, apply the model many, many times and measure the errors.

DR. LACKEY: We work in large measure with stochastic models and to a much lesser extent with deterministic ones. I would agree with the comment that there is more in the distribution of your estimate than in the point estimates. I think that this is often the way to go because the evolution of models does not necessarily go from deterministic to stochastic.

DR. KNEESE: When we did a little simulation work on the Potomac River a few years ago using a stochastic streamflow generator and looking at a reservoir that the Corps had proposed for low flow augmentation and water quality improvement, it appeared that changing the allowable probability of violation of the DO standards could have as large an impact on the system cost as, let us say, keeping the violation probability constant and dropping the DO standard from 4 parts to 3 parts per million.[1] That is another dimension of standards, and the question is, how much can models help us in taking that kind of decision?

---

[1] Robert K. Davis, The Range of Choice in Water Management, (Baltimore: Johns Hopkins University Press for Resources for the Future, 1968.)

DR. THOMANN: There has been a fair amount of work on the inclusion of probabilistic statements in DO models, but I think we are not at that point yet in most of the ecological models.

DR. O'CONNOR: I think your comment is well taken. We need a few more years of experience in this.

DR. THOMANN: Tell that to our decision makers.

DR. NIHOUL: I want to answer your question about statistics and dispersion. We do try to take this into account by predicting upper and lower bounds. In any case, the mean value is going to be significant in the same sense in ecology as it is in hydrodynamics. You know when you measure a current with a current meter, that you don't get the turbulence, but however essential the turbulence is for the dispersion, the mean current given by the current meter has a useful physical meaning. Obviously, the same must be true in ecology. Mean values are meaningful. You must, however, try to complete the picture with upper and lower limits.

DAY TWO-- MORNING

O'CONNOR, DiTORO, THOMANN PAPER: "The Lake Erie Model...and Others"

* * *

DISCUSSANT: HARRISS

* * *

DR. HARRISS: I have some real concerns about the basic policies
that are controlling the directions in which we are going--scientifically
and technologically--and what some of the consequences of these might
be. The hypothesis around which I am going to work is that in most of
these models we are solving short term problems. We are thus using a
philosophy which, I think, is misleading because we are not going to
protect the quality of environment in the long run. Rather, we have a
scientific system which is being designed to maintain our existing economic
system.

Let me talk specifically about my concerns as far as the existing
modeling attempts go. First, let me address those which are characterized
as ecosystem models.

To assess these models I want to look at three general criteria:
First, how well they are identifying energy-flow pathways both in time
and space? Second, how well they are able to handle mineral fluxes
in time and space? And, third, how successful they have been in identifying
and using so-called "unique function" organisms? (That is, organisms which
make a unique contribution but which have significance neither in the
energy nor the mineral flow in the system. I think this is a very
common occurrence, is of considerable importance in trying to assess
whether the ecosystem models will be successful.)

I would suggest that most ecosystem models, as of this date, have failed to contribute to a better understanding of the basic processes related to the energy-flow or to the mineral cycle. Some ecosystem models are to the point where they provide reasonable descriptions of what we see today in the ecosystem. However, most of the things that we see today are only the coarse phenomena. For example, we are getting rather good and reasonable descriptions of acute cultural eutrophication in Lake Washington, Lake Erie, and the like. But I believe that we are really getting into a closed circle. In other words, we are developing models which verify effects which are more or less obvious to the biologist to begin with.

With respect to the ecosystem models, pollution biology and aquatic resource management, the real problems I see are the long term sublethal effects of very small quantities of pollutants and the very small changes in mineral fluxes which can cause a change in the ecosystem.

I think that with a few examples I can illustrate that these can be major effects. Even though we may not have techniques sensitive enough to detect these chronic effects now, in the long run they will almost certainly appear as major problems to be dealt with.

For example, consider the behavioral effects of sublethal exposure to pesticides or mercury. In these cases you don't see large buildups of mercury in the environment, so you don't spot these problems by going out and setting up a monitoring system. These are changes that can occur due to something like the reduction in photosynthesis in one or two critical species. For example, you might get a slow, selective elimination of certain nannoplankton, so that a food-web is disrupted.

An even more critical problem--and one that is well documented
scientifically--is the development of resistant populations of organisms.
These are populations exposed to very slight but long-term increased
exposure to a contaminant. Very gradually, through genetic changes, the
population develops resistance to these contaminant exposures. The
resistant organisms can carry extremely large concentrations of the contaminant and are lethal to non-resistant predators. The effects on
the food chain, which have been demonstrated particularly for pesticides,
can be disastrous.

Another general problem area is identifying unique organisms, which
are the key to the functioning of the whole system, and yet, which appear
to have very little significance as far as energy and mineral flux are
concerned. A few examples might help here.

In the Florida Everglades ecosystem, the alligator plays a critical
role in the survival of the whole system. Alligators, from a biomass
standpoint or from a mineral cycling standpoint, have a negligible role.
But, it turns out that during the dry season the alligator likes to dig
holes and stay down below the water table. The other aquatic organisms
will migrate to these alligator holes during the dry season. These are
very small sink holes, but not only do they feed the alligator, which is
of very practical importance to that particular organism, but they also
provide the aquatic breeding stock when the system revives during the
wet season.

Too many of the ecosystem models are so aggregated that such a function is totally left out. The same thing holds true for certain bird
species in tropical rain forest ecosystems. In many cases there are

examples of birds which are critical for seed dispersal. The whole
system revolves around these seed dispersal mechanisms, and yet these
birds are not important as far as the mineral or energy flow are concerned.

Very briefly, the question is, Can ecosystem models be made complex
enough to handle these problems? I am an optimist by nature, but I
doubt it. I think that the problems just discussed here, particularly
genetic problems and the problems of the unique species functions, are
probably beyond the capability of most models. I am afraid the ecosystem will be fairly well deteriorated by the time the models get
sophisticated enough to predict chronic impacts.

But, I believe that the models have played and will continue to
play an extremely interesting role as far as organization of research
goes.

Now, I will comment briefly on the limitations of what I call the
applied models or the models aimed at specific problems. First, we all
agree that there is a real problem with identifying environmental quality.
We can take dissolved oxygen but, in many cases, by focusing on the
DO problem, we are simply providing a healthy environment in which to
breed nasty mutations, because while we maintain the oxygen level, an
industry happens to be releasing organic or metallic contaminants which
get into the population at very low rates causing the problems discussed
previously. In my experience no waste discharge is simply a dissolved
oxygen problem. A model treating oxygen is only a partial solution at
best and could lead to more severe problems in the long term.

Second, I am concerned that the important problems today are not
the point sources. I think the real problems are the non-point sources

such as agricultural waste and storm water runoff. These are problems
which to me pose very difficult modeling questions, and I am not sure
that the models will meet the challenge.

DAY TWO - AFTERNOON

SCHAAF PAPER: "Fisheries Models: Potential and Actual Links to Ecological Models"

(Comments subsequent to paper but prior to formal discussion.)

DR. NIHOUL: The equations which you show on page 13 of your paper are basically modified Lotka-Volterra equations. Do I understant that K_1 is taken as a constant?

$$\frac{dB}{dt} = BK_1 \, (L - B) - QBE \qquad \frac{dE}{dt} = K_2 E \, (B - b)$$

DR. SCHAAF: Yes.

DR. NIHOUL: Presumably this is a function of the food which is available to fish. Do you know anybody who has tried to simulate these equations with K_1 being given as a function of time, that is, a periodic or random function?

DR. SCHAAF: I don't know of anybody who has tried to do that.

DR. NIHOUL: We did something with K_1 as a function of time in Lotka-Volterra equations. The results we got were really quite surprising. You can get parametric oscillations in the system, leading sometimes to the extinction of one of the species. If you take K_1 as a random function of time, then you also get some surprising results. I would like to send them to you in order to have your comments. I don't know if they area realistic.

DR. SCHAAF: Please do. I know of nobody who has really looked at K_1 in this manner. It relates to the problem I was referring to of the boundaries of our system, because, obviously, the initial conditions determine the fate of the system in that kind of formulation.

* * *

* * *

(General discussion.)

DR. ORLOB: I would like to ask a question of Dr. Schaaf. One of your equations has a little "e" for environment at the end. That is the expression that really interests me, of course, because that is where pollution and environmental changes induced by man impinge on fisheries. Who in the fisheries modeling field is concerned with that term? Is anyone looking specifically at it and trying to put quantitative environmental variables into the actual models?

DR. SCHAAF: I know of nobody in the marine fisheries field who is looking specifically at that problem. Perhaps some of the fresh-water fishery biologists and recreational fishery biologists are more immediately concerned with it because the impingement is a little more visible. In most marine fishery situations--particularly the commercially important species--the impingement is not very evident now. In these fisheries fishing mortality rate comprises a large fraction of the total mortality rate. It is the thing that can be manipulated most directly and most immediately. Therefore, the initial attempts at modeling marine fisheries focused on it. We don't know what to do about the environmental problems for immediate purposes, and, in my opinion, they are not as directly relevant to fisheries management right now as some of these people-management problems we are faced with.

DR. LACKEY: I would take the same information and look at it from a different standpoint. I would allocate the impact of pollution to the first elements (the vital statistics) of that equation. In other words, if you introduce   thermal pollution, I would say that its effects might be allocated to the recruitment coefficient, or one or all of the other vital statistics. I would regard the "e" as a stochastic parameter. Any pollution effects should be accounted for in the first four coefficients, with random shocks falling into the error or stochastic term.

DR. JOEL O'CONNOR: With regard to Dr. Orlob's question, we are now gathering data appropriate for fisheries models. We are not developing models but we are looking at some of these error terms. Within the New York Bight there areseveral stages. We are looking at effects on egg and larva mortality from contaminants, and at the impacts of toxic sediments and contaminants in the water column itself.

DR. ORLOB: But, do these same things really fit into the framework of longer range planning? Dr. Schaaf mentioned that the fisheries model usually addressed a short term crisis situation. Someone, who spoke earlier, was worried about the long term chronic effects. How do we begin to include these? If we are going to include them, we begin to face the problem of projecting--even if we are not entirely on solid ground.

The ultimate demise of fisheries may come about for reasons that are inadvertently ignored, if we address only today's acute problems.

DR. O'CONNOR: Right. That is our concern; that long-term lethal and sublethal effects, particularly in the early life-history stages. In our case the concern is not for individual species. I doubt that the MESA project will be involved with the classic population dynamics studies.

DR. ORLOB: If there is not a consensus, there is at least a strong feeling that we are on shaky ground in trying to deal with long-term chronic effects using ecologic models.

DR. KNEESE: I would like to follow up on Gerry's (Orlob) question about pollution and its role in fishery modeling. Is anything significant along the modeling line being done with shell fisheries?

DR. SCHAAF: I don't know.

DR. CHEN: We tried it, but it did not work.

Now, I come to my question. In the marine fishery field I think there is great concern about the amount of fresh water being brought into estuary systems. The understanding is that marine fisheries require fresh water to bring in food. In that context, what kind of impact would zero discharge have on a marine fishery?

DR. SCHAAF: I can imagine it would have a deleterious impact.

DR. CHEN: So, if you were to develop a reservoir on a river system and modify the hydrologic regime, how would that affect the fish?

DR. LACKEY: The example we use in management classes is the Aswan Dam. I don't know if anyone did any modeling on that.

MR. CROOK: Yes, they did.

DR. LACKEY: The shrimp situation in the Gulf of Mexico is another example. It is very sensitive to turbidity.

DAY TWO - AFTERNOON

O'NEILL PAPER: "Managing the Modelers"

* * *

DISCUSSANT: CROOK

* * *

MR. CROOK: I want to compliment Bob (O'Neill) for a very interesting paper. He said a lot of things that I wish I had said before
he did. He said them well.

I think Bob (O'Neill) has covered the topic as I can't and will
not try to. However, I do have some concerns about the different value
systems of the researcher, the modeler, the decision-maker, and the
planner. These reward systems are so different that it is hard to keep
them all in line. I think Bob (O'Neill) has given us some ideas as to
how this might be done.

Certainly, if you can merge these people into a team, that is a
pretty good way to handle some things. But I am concerned that you
may lose management control if you give the team too much leeway. The
team has a motivation of its own. I am for relatively strong management control of operations of this type. I don't believe in turning
modelers loose to solve my problems. I believe the successful modeling
efforts, which I know about, almost always have happened as a consequence of relatively strong management control and involvement over a
continuing period of the modeling development and application.

I don't separate development and application. We have heard
something about this, but really, you ought to make this a part of the

entire project, that is, that you not only test it but you apply it and use it. In other words, the people who have helped to develop the model will have to make it work to solve real problems. If you do this, you will build in a responsibility that will help you in many ways.

We were asked to state our opinion as to where we are in ecological modeling. Frankly, we are really not very far along--as many of you have already acknowledged. However, that does not mean that what we know isn't useful. What you can develop is better than what the decisionmaker probably has without your effort.

I think that Hydroscience's work for us had paid us off in a number of different ways already. We are developing a model that can be expanded, can be used for more than planning purposes, will be refined, and may be utilized in increasing intensity in smaller and smaller scale, down to local situations, as we accumulate more knowledge and ability.

In connection with structuring a model for management purposes, what are the inputs? You ought to start working with modelers and scientists and users of models from the start. This means that the inputs will be, initially, from available data. Look at these, and see if you can design a model. It may surprise you. You may be able to get a simple model and come up with some reasonable answers that management likes. Then, once you have done something for management and told him some answers, you are in a much better position to say: "Hey, I need a couple of million for this kind of data collection; how about it?"

He, most likely, will oblige because he now has confidence in you.

But, I do not buy the story that you ought to collect only data to suit your model. There are large quantities of data that the Geological Survey, NOAA, and the Weather Bureau have been collecting for a long time. But every time we go back into history to get additional information to run a model or solve a problem, we find out that not enough data were collected in enough different areas on enough different parameters. That does not, however, mean that you have to collect an unlimited amount of data on every conceivable aspect of the environment.

You should have as simple a space relationship as you can because of the cost factors. Leave it to discovery to develop what additional complexities you need. The same thing holds true with respect to time. Make it as simple and as long a period of time as you think you can get by with. As Carl Chen said, sometimes your space and time relationships can be very general and long-term and still be quite rewarding. (Whether you need a steady-state model or you want to look at transient effects or get all the peaks and valleys, depends on the problem.)

How accurate do your models need to be? They need not be one bit more accurate than necessary to give you the right answer. That says a lot, really. Many of you people want to build the ultimate into your model. I don't want the ultimate. I need an answer, and I need it sooner than you can give it to me and at less cost than you are going to charge me. So, let's keep it simple. Now, there may be on the research side--and I said I was talking about planning--every reason to do something in great detail with a lot of refinement, but I don't want

you to bill me for it. I only want to pay for the minimum amount.
I want to use your expertise.

If I were attempting a new project, I would try to obtain--as
Bob (O'Neill) says in his paper--the most competent individuals to
work with me. They would have a proven track-record. They would be
productive. I would put most of my money behind them. If it were
necessary, I would separate from the proven, productive group the innovative, hard-to-work-with scientist who comes out with earthshaking
discoveries once or twice in his lifetime. I would encourage him to
produce in a proper climate. You don't have to buy too much of this,
because the university and its set of rewards takes care of these people.

I would take a young, competent, dedicated person who knew the
planning and management side well and who had some experience with modelers
and models. He would be in charge of my program. I would ask for a man
who had the credentials in his own field and who had the attitudes that
Bob (O'Neill) outlined. I would let him work on this, and I would trust
him--with constant reporting--to handle it. I would give him centralized
funding control so he could pass out the money on the basis of productivity.

He would start simple, but he would start. He wouldn't delay the
model until he got all the data and tested out all the hypotheses and
went through all the research efforts necessary. He would start modeling
early. He would progress from the central issues and the interrelationships to the refinements. He would involve decision-makers with repeti-
tive, early reports, which would be status reports--not necessarily

answer-producing reports.

I would try to test some of the early results in other disasso-
ciated programs of a similar nature. Then, I would leave a good deal
of latitude for error. I would hold back some money which I wouldn't
even give to the young manager. You are not going to go through this
operation smoothly and easily. You are going to run into hangups,
and you are going to have to have the resources to take care of them or
you fail for want of just a little more effort.

DR. KNEESE: I would like to hear Jacques (Nihoul) say something
about the management of the North Sea modeling work. It involved not
only different disciplines and institutions but also different national-
ities, some of which are not overly fond of each other.

DR. NIHOUL: There was a group formed called JONSIS (Joint North
Sea Information System). This group addressed itself to the various
countries situated around the North Sea. On an informal basis they
invited experts into a joint campaign for oceanographic cruises and,
since last year, joint modeling efforts.

In the JONSIS context there are groups which are formed to organize the experimental surveys - Joint North Sea Data Acquisition
Programs (JONSDAP). The composition of these groups may be different
from one experiment to another.

Money is provided by each country. The national experts will
commit vessels and equipments according to their countries potential
and subject to approval by their respective government. There is no
money in JONSIS itself.

It is the same situation with the International Council for the Exploration of the Sea to which JONSIS reports its activities. These are merely representatives supported by their own countries.

DR. KNEESE: On the surface it seems you have in the North Sea operation both a successful program - you actually did model systems and you actually predicted - and a distinct lack of overlying management structure. Am I mistaken about that?

DR. NIHOUL: As I say, JONSIS is informal. The people who are working in JONSIS are supported by their own governments. There is no "head" of JONSIS. We change the chairman at each session.

MR. CROOK: What are your objectives?

DR. NIHOUL: Our objective is a better understanding of the North Sea. This will be attained, we hope, by getting the necessary data and building the necessary models (storm surges and tidal models, residual current models, ecosystems models, etc.).

DR. CROOK: Could the cost, time and effort have been reduced had you had central management?

DR. NIHOUL: I don't think so, because with so many different countries involved it has to be a gentleman's agreement. It is not like the separate states in the U.S.

DR. CROOK: There is some similarity.

DR. NIHOUL: I have a feeling that this international program can only work with a gentleman's agreement between people who can talk in the name of their country. These must be people who can go back to their respective countries and, since they are in charge of national programs, will be listened to by their governments.

DR. KNEESE: What kept it together? Was it the mutual interest of the scientists? How could you be sure that a piece of information would be available at a certain time to help the model?

DR. NIHOUL: We plan our own experimental service in JONSDAP. We are sure that the data acquired in this way will be valuable. The people who make the decisions in the JONSIS meetings plan the data surveys as much as they do the models. There are problems, of course.

MR. CROOK: I suppose international problems of this nature are similar to those within a set of universities or states. Andy (Robertson) has some experience with an international data collection program: IFYGL. There were many problems which could have been ameliorated--if not solved--by having centralized control of funding and personnel. What do you think, Andy (Robertson)?

DR. ROBERTSON: I would agree that centralized funding and control of the personnel probably would have given us more progress towards the objectives which the planners and directors had set up.

DR. NIHOUL: Perhaps my answer was not quite clear. JONSIS is working by gentlemen's agreement, but, whenever we decide to do some-thing--for example, a hugh experimental survey in 1976--we set up a group to organize it. The group has a chairman who is responsible. When JONSIS decided to make joint simulations of problems in the North Sea, they set up a group, JONSMOD (Joint North Sea Modeling Committee) to prepare the 1976 campaign and future campaigns. So, there is this kind of breakdown whenever there is a definite objective.

DR. ROBERTSON: JONSIS seems to have done one thing that might have worked in IFYGL. Often, when we were cooperating on a project, we

designated one American and one Canadian to oversee the work. Sometimes that worked well; other times not so well. Maybe JONSIS had
an advantage, in that with so many countries involved you could not
pick co-chairmen and so opted for having one person with overall
responsibility.

DR. THOMAS: Were you able to sample in Canadian waters at will?

DR. ROBERTSON: Yes. I don't believe on the Great Lakes there
has ever been any problem on sampling the waters of the other country.
There are problems in getting into foreign ports, sometimes.

DR. THOMAS: In the North Sea could you easily sample other terri-
torial waters? I gathered from what you had said that you could not.

DR. NIHOUL: Yes, we can, but we have to ask permission. There
is no problem. However, if we have two vessels--one Belgian and one
British--and we want to survey the southern part of the southern bight
between Belgium and England, it is certainly cheaper and simpler to
have the British vessel survey the British region and the Belgian one
survey the Belgian region.

DR. THOMAS: Do you put your scientists on a British vessel?

DR. NIHOUL: Certainly. We do this all the time.

DR. ROBERTSON: I can cite a problem of this kind. Both we and
the Canadians put out buoys to measure current and temperature--each
country puts its buoys on its side of the lake. For comparison, we
put one pair next to each other. As you might expect, there were some
differences between the readings. It would have been better to have
interspaced the buoys.

DR. NIHOUL: We have an inter-calibration program. Each country makes the same measurements and analysis with its own techniques, and the figures are compared--not everywhere, but at a few points.

DR. RUSSELL: We have been talking about managing modelers when they are all biologists. There may be some variation in specialty, but, to a large extent, they speak a common language.

We are also talking about modeling in a management context. That implies--at least for me--the presence of other disciplines quite different from biology and hydrodynamics; for example, economics and other social sciences. I think this introduces an order of magnitude greater problem of creating teams which can actually do something useful.

I have seen two kinds of results from such attempts; neither of which constitute a truly interdisciplinary management model. One is the book in which the biologist writes a chapter, the economist writes a chapter, the engineer writes a chapter, and so forth. The product, then, is a disconnected set of narrow papers. The other result is the imposition of one discipline's value structure on the entire project. From what I have seen, this tends to be correlated with the "great man." Every part of the research or the management effort may take on the value structure which comes out of his discipline.

DR. CROOK: In our efforts to model the Great Lakes we chose to take the physical, chemical, and biological systems separately and to add exogenously the social sciences, the economics, the population dynamics, and the loadings that came from these sources. We did the latter in a separate way because we could play them one against the other. We add these other things in as plug-in's   It is the blackbox approach, if you will.

DR. RUSSELL: You have to know that the plugs fit, however. You don't want a three-pronged plug coming out of the economics box to be put into a two-pronged socket in the biological box. We have faced this problem. It may seem silly, but the only data around on organic material in industry discharges, at least prior to the new permit system, was for BOD-5. An ecologist, I believe, finds BOD-5 unsatisfactory as a measure of organic load. But something has to be worked out, otherwise separately developed economic and ecological models will not work together. And if they don't work together, you don't have a management model.

DR. ROBERTSON: We had exactly that experience. We tried to get a conversion factor from BOD to organic carbon, and it didn't work very well.

DR. DONALD O'CONNOR: That information is around. The industrial discharge people know what is in there. They know because they are putting in the raw products. I think they will have to be taught a little bit of responsibility at some point along the line. In this sense Bob Harriss' remarks were pertinent. He said we are only looking at those models which are functional, whereas the things we should be looking at are the toxic substances and synthetic chemicals which are coming in.

DAY THREE - AFTERNOON

ORLOB PAPER: "Concluding Observations: Problems, Prospects,
and Promises"

(Interruptions concerning a conceptual model schema discussed
by Orlob.)

DR. DONALD O'CONNOR: Where do you introduce the toxicity? In
the growth term?

DR. ORLOB: It is a constraint on growth. Carl (Chen) has
introduced it as a modification of mortality rate in the net growth term
in phytoplankton. For example, you might use it as a modifier of the
natural mortality rate, increasing that rate if the toxic influence
were above a threshhold that may be considered tolerable by the organism.

DR. DONALD O'CONNOR: So, you are putting it into the death
rate rather than the growth rate?

DR. ORLOB: Yes.

DR. CHEN: I feel that the effects of toxics ought to be included
in both places.

DR. SCHAAF: Why did you have what appeared to be a more complicated schematic for zooplankton than you did for phytoplankton?

DR. ORLOB: There is added detail in the case of phytoplankton as
well, but the decision has been made that phytoplankton be considered
as a total group rather than a subgroups. The zooplankton were divided
primarily because of their influence on the fish stocks which were a
major concern. Then, of course, there was the recognition of the
temporary nature of certain zooplankton stocks.

DR. WOLFE: You showed the phytoplankton and the zooplankton elements of the system. Are there analygous subsystems all the way up to predation by humans?

DR. ORLOB: Predation by humans is, of course, implicit, but there is no fisheries management model as such. That, of course, should be developed at some point.

DR. WOLFE: How about transport of some of the toxic materials in the system through fisheries products back to humans. This is one of the major points at which management of toxic materials and management of water quality usually occurs.

DR. ORLOB: At the present time the questions of toxification are looked on as primarily restraints on growth or additions to mortality. I cannot say at this time whether or not the fate of the toxicant per se--through the food chain in terms of mass and concentration--is envisioned as completely as is needed in order to calculate the loads getting back to man through his harvest of marine organisms.

DR. WOLFE: The pathways for those flows may in fact be similar to or identical with the pathways of carbon flow which are being monitored.

DR. DONALD O'CONNOR: I would subscribe strongly, at this conceptual stage, to the notion that the buildup of toxic materials in the food chain be considered and recorded in much the way you did it here.

DR. WOLFE: The output of the toxicity submodel apparently was introduced both to the growth-respiration ratio and also to the natural mortality rate. The development of that output appeared to include the time dependent accumulation of the toxicant in the component.

DR. ORLOB: That is the case with certain heavy metals.

DR. JOEL O'CONNOR: One thing I should stress, which would make ecologists feel a little easier, is that the flow diagrams with this level of complexity, that is, the ones you have put up here, are developed for the conceptual model only. It is clear that we are not going to implement an entire model like this at this level of complexity.

One of the questions, which we are to answer very soon, is: What parts of this can we model in a quantitative sense and what parts of the model will perhaps have to be implemented at some higher level of aggregation?

DR. ORLOB: That is why I showed you three levels at which zooplankton appeared. There was one aggregate compartment, then two subcompartments, and then various stages of maturation in great detail. This at least provides us with a vehicle for a better understanding of what we are sacrificing when we do make that simplification.

DR. WOLFE: May I suggest that, in addition to bacteria, you also include fungi and protozoa, at least conceptually?

DR. ORLOB: I am willing. I simply don't have the ecologic background to be able to make that kind of recommendation with confidence. I do feel, however, that there is evidence--from what I have seen here at this symposium--to suggest that this would be a very important step forward.

DR. KELLY: I would like to reinforce something that you glossed over in your recommendations, that is, that we take a closer look at the benthos. The sediment - water interaction - as far as I know is very poorly understood. A lot of work needs to be done there.

DR. NIHOUL: I would like to comment about what was said in connection with the importance of the benthos. There is now a NATO Conference on the benthic boundary layer. It includes biologists, chemists, sedimentologists, and hydrodynamicists. There are 60 people. A book will be published. You might be interested in seeing it.

DR. ORLOB: May I add, along that line, that late last spring there was a conference in Ottawa, Canada, on the persistence of chemicals in the aquatic environment, including marine systems. It included people from the marine bacteriology areas. An excellent paper was presented by Dr. Caldwell of the University of Maryland on the experience in Chesapeake Bay. The research reported was directed to the bacterial problem and to acute and chronic toxicity considerations.

DR. DONALD O'CONNOR: We should concern ourselves with those areas that bring us together. Bob (O'Neill) said in one of his comments that we begin by talking at each other and then ultimately talk with each other. Let's find out what those elements are that do pull us together so that we can define a better scientific profession structure for peer review and analysis of all the problems.

My final point is this. When we do get together, we must remember that we come from two widely separated ends of the spectrum. One person looks at the world and says: "There is order; I can quantify it." This is the Newtonian heritage. The people who do the mathematics come from that philosophic heritage. There is also, however, the Aristotelean heritage. When you look at the world, you see its

light and its beauty. Why is it? It is so because of its diversity,
its complexity, and its stochastic nature. You can never make order
out of it.

Somehow we have to accept both of these. Both are necessary.
I think it is the latter view that really comprehends the world.
What we try to do is to try to make it somewhat predictable and usable.

DR. BROWN: Let me get into something different. I was particularly interested in your recommendation, Gerry (Orlob),which suggested
spending more effort looking at non-equilibrium situations. This
seems to me to be particularly important as we try to fit the model
into intermediate assessments of what is going to happen in the
environment.

It seems to me that we might consider trying to take advantage of
catastrophies. We can think about being ready with our model and going
in and actually trying to apply it in the catastrophic situations--
hopefully using some data and simulating just what happened in that
extreme condition.

DR. DONALD O'CONNOR: I have a specific suggestion for NOAA or
RFF. I would suggest sponsoring a session at which a combined group
of the quantitative and qualitative scientists--in conjunction with
both Federal and private people--look at the re-education of the
future scientist-engineer who is going to deal with natural resource
problems.

One necessary element in this grouping would be the representation of the spectrum that I mentioned, that is, the ecologist and
the engineer.

This group should support and structure, at some of the better universities, programs which would bring together faculty with the diversity I've talked about. I don't know whether we can do it, but, if we can, the next generation will be trained--and they will do the job.

It is a long-range suggestion, and it seems to be a governmental responsibility because natural resources fit better into the public domain.

DR. ROBERTSON: Haven't a great number of universities tried to do this? I was on a committee that had civil engineers, botanists, and zoologists. Our mission as faculty members was to set up an ecology institute. Nothing much came of our efforts.

Some interdisciplinary environmental institutes have been set up. I have not really seen the results from those. I do know that a lot of them have not gone very well.

KNEESE AND RUSSELL CLOSING REMARKS

DR. KNEESE: When I said that Gerry (Orlob) was going to summarize, integrate, and interpret, I was being facetious. But he really did it. And, he did it exceedingly well, in my opinion.

I want to say, too, that I have enjoyed and appreciated the conference. I don't say that about many conferences. It seemed to me it was an instance where people came together who really needed to talk to each other.

DR. RUSSELL: I think anything I could say would be anti-climactic. But if I had to summarize in the quickest possible way what I have learned, I would say that I got an affirmative answer to my question about convergence. I also asked about the state of the models. I now have a strong feeling that the models are considered pretty good up to phytoplankton and not much beyond that. I asked questions about the management context and I have the impression that this is where we really need to do a lot more work together. Even though the people here were chosen--as far as we could do it--because of an expressed interest in real problems, there is a long way to go before we are really talking the same language about management.

I would like to thank you all for coming. For me, and I hope for you, it has been most educational.

For Product Safety Concerns and Information please contact our EU
representative GPSR@taylorandfrancis.com
Taylor & Francis Verlag GmbH, Kaufingerstraße 24, 80331 München, Germany

www.ingramcontent.com/pod-product-compliance
Lightning Source LLC
Chambersburg PA
CBHW060750220326
41598CB00022B/2391